TRACE ELEMENTS IN MAGMAS
A Theoretical Treatment

Studying the distribution of certain elements, present in very low concentrations in igneous and metamorphic rocks, can yield important clues about the rocks' origin and evolution. Trace elements do not give rise to characteristic minerals, but their behaviour can be modelled to provide historical information about the source magma. This book brings together the essential theory required to understand the behaviour of trace elements in magmas, and magma-derived rocks. It presents a wide range of models and mechanisms which explain trace element distribution.

Trace Elements in Magmas provides an excellent resource for graduate students, petrologists, geochemists and mineralogists, as well as researchers in geophysics and materials science.

DENIS SHAW joined the Department of Geology at McMaster University, Ontario, in 1949, continuing his research as Professor Emeritus until 2003. Throughout his career, he taught courses in geochemistry in Canada, France and Switzerland. He worked as editor for a range of publications including *Geochimica et cosmochimica acta* and the *Handbook of Chemistry*. In 1964 he served as President of the Mineralogical Association of Canada, and received the Distinguished Service Award of the Geochemical Society in 2002. Professor Shaw passed away in October 2003 and his widow, Susan Evans Shaw, and Cambridge University Press are grateful to Professor Stuart Ross Taylor for his work in editing the final manuscript in preparation for publication.

STUART ROSS TAYLOR, a trace element geochemist, is an emeritus professor at the Australian National Univeristy and is the author of *Solar System Evolution: A New Perspective* (Cambridge Univeristy Press) and several other books.

T0211333

CAMBRIDGE UNIVERSITY PRESS
Cambridge, New York, Melbourne, Madrid, Cape Town, Singapore, São Paulo

Cambridge University Press
The Edinburgh Building, Cambridge CB2 8RU, UK

Published in the United States of America by Cambridge University Press, New York

www.cambridge.org
Information on this title: www.cambridge.org/9780521822145

© S. Evans Shaw 2006

This publication is in copyright. Subject to statutory exception
and to the provisions of relevant collective licensing agreements,
no reproduction of any part may take place without the written
permission of Cambridge University Press.

First published 2006
This digitally printed version 2007

A catalogue record for this publication is available from the British Library

ISBN 978-0-521-82214-5 hardback
ISBN 978-0-521-03634-4 paperback

Cambridge University Press has no responsibility for the persistence or accuracy of URLs for external
or third-party internet websites referred to in this publication, and does not guarantee that any content
on such websites is, or will remain, accurate or appropriate.

TRACE ELEMENTS IN MAGMAS

A Theoretical Treatment

DENIS M. SHAW

Formerly of McMaster University, Ontario

Edited for publication by

STUART ROSS TAYLOR

CAMBRIDGE
UNIVERSITY PRESS

Geochemistry is a compilation of imprecise, irreproducible and uncoordinated analyses.

(i) Keep the rocks in mind, for they cannot be reduced to analytical measurements; (ii) (from O. F. Tuttle) minerals are the archives of the rocks; (iii) keep filing [your] fingernails while waving [your] arms.

Contents

Preface   *page* xi

1 Introduction   1
   1.1   Defining trace elements   1
   1.2   The quality of trace element data   3
   1.3   Sample heterogeneity   5
   1.4   Mineral analysis for trace elements   9
      1.4.1   Sampling   9
      1.4.2   Chemical analysis   10
   1.5   Summary   12
        References   12

2 Partition coefficients   14
   2.1   Introduction   14
   2.2   Solutions with a common solute   15
   2.3   Reacting solutions: law of mass action   17
   2.4   Effects of variation of temperature and pressure   19
   2.5   Measurement of partition coefficients   23
   2.6   Extended theory   25
   2.7   Major element effects   27
      2.7.1   Olivine–melt partition   27
      2.7.2   Redox effects   29
      2.7.3   Volatile fluids   31
      2.7.4   Melt structure effects   34
   2.8   Influence of the host solid   34
      2.8.1   Substitution site deformation   34
      2.8.2   Influence of mineral chemistry   39

	2.9	The Henry's law controversy	43
	2.10	Use of partition coefficients	44
	2.11	Summary	45
		References	46
3	Crystallisation: basic trace element modelling		51
	3.1	Introduction	51
	3.2	Definitions	52
	3.3	Temporal variables in a crystallising system	52
	3.4	Equilibrium crystallisation	53
	3.5	Fractional crystallisation	55
	3.6	Mineral zonation	61
	3.7	Intercumulus trapped melt	63
	3.8	Mineral pairs	66
	3.9	Incremental or stepped crystallisation	66
		3.9.1 Constant melt proportion	68
		3.9.2 Constant mass increments	70
	3.10	Summary	72
		References	72
4	Crystallisation: variation of mineral proportions, partition coefficients and fluid phase proportion		74
	4.1	Introduction	74
	4.2	Variation in mineral proportions	75
	4.3	Variation in partition coefficients	78
		4.3.1 Trace elements	78
		4.3.2 Major elements	83
	4.4	Crystallisation in the presence of a fluid phase	86
		4.4.1 Instantaneous degassing	87
		4.4.2 Continued fluid release	89
		4.4.3 Discussion	90
	4.5	Summary	92
		References	92
5	Crystallisation assimilation, recharge and eruption		94
	5.1	Introduction	94
	5.2	Resorption or assimilation	94
	5.3	Mass balance	95
	5.4	Assimilation by melting and solution	99
	5.5	Assimilation by reaction	103

5.6	Assimilation-fractional crystallisation processes	105
5.7	Magma recharge and discharge	105
	5.7.1 Conservation of initial magma mass	106
	C+A → E → R	108
	C+A → R → E	111
	5.7.2 Conservation of residual magma	112
	C+A → E → R	112
	C+A → R → E	113
	5.7.3 Discussion	113
5.8	Recharge, eruption, assimilation: the rate process model	116
	5.8.1 Conservation of the initial magma mass	118
	5.8.2 Magma mass M is not constrained	119
5.9	Summary	122
	References	123

6	Trace element evidence for crystallisation processes	124
6.1	Introduction	124
6.2	Variation diagrams	124
6.3	Other two-element plots	126
	6.3.1 Crystallisation trends	129
	6.3.2 Element ratio plots	133
	6.3.3 Mixing and assimilation	134
6.4	Inversion modelling	135
6.5	Summary	139
	References	140

7	Melting: basic trace element modelling	142
7.1	Introduction	142
7.2	Melting a heterogeneous rock	142
7.3	Early partial melting	148
7.4	Definitions	149
7.5	Bulk partition coefficient	150
7.6	Trace elements in equilibrium melting	150
7.7	Trace elements in fractional melting	152
7.8	Modal and non-modal melting again	158
7.9	Incremental batch melting	160
7.10	Batch melting with retained melt	163
7.11	Equilibrium melting vs. fractional melting	168
7.12	Melting in the presence of volatiles	170
7.13	Disequilibrium melting	179

7.14   Accessory minerals entrained during melting   181
7.15   Summary   183
       References   184

8   Melting: more complex processes   187
    8.1   Introduction   187
    8.2   Incongruent and reaction melting   187
          8.2.1   Simple incongruent melting   188
          8.2.2   Reactive melting   194
          8.2.3   Three reacting phases plus an inert phase   197
          8.2.4   More complex reactions   199
    8.3   Variations in mineral proportions and partition coefficients   201
          8.3.1   Variation in mineral proportions   202
          8.3.2   Variation in partition coefficients   202
    8.4   Rock melting by zone refining   205
    8.5   Summary   208
          References   209

9   Dynamic mantle melting   211
    9.1   Introduction   211
    9.2   Dynamic melting   211
    9.3   Closed system model   215
    9.4   Open system model   220
    9.5   Discussion of models   227
    9.6   Melt dynamics   232
          9.6.1   One-dimensional motion   232
          9.6.2   Two-dimensional motion   233
          9.6.3   Percolation of melt through mantle   237
    9.7   Summary   239
          References   240

    Index   242

Preface

The years following World War II saw a steady improvement in the analysis of rocks for minor and trace constituents. It became clearer that trace elements were not haphazardly distributed and that chance played a minor role. To find the principles of distribution of the elements was one of the aims of geochemistry, according to V. M. Goldschmidt, and researchers began to try to understand trace element behaviour.

Two main approaches developed: one was aqueous geochemistry, where the emphasis was on the oceans and mineral genesis reactions in electrolytic solutions; the other studied the igneous and metamorphic rocks and, to some extent, metallic deposits. The second, often inappropriately called *hard-rock* or *solid-state geochemistry* was, like the former, concerned with heterogeneous phase reactions, but of different kinds. Central to it is the concept of the *partition coefficient*, and much effort has been expended in attempts to measure such parameters or variables. Many schemes were proposed and tested to show how some observed trace element or isotopic distribution pattern could be explained in petrological terms, and such accounts are scattered throughout the literature, camouflaged under various titles.

This book constitutes an attempt to gather together the wide variety of possible models or mechanisms to explain the distributions of trace elements in igneous, metamorphic and metasomatic rocks, so that they are available for application as needed. The emphasis has been quite deliberately placed on the details of the mechanisms and, as a consequence, few examples have been cited.

Another reason for the paucity here of examples from the natural world is that there are, up to the present, few trace element studies available where the analytical precision is sufficient to choose among different models, although this is not the case with many isotopic systems.

Much of the material here has formed part of graduate courses and I am grateful to successive waves of graduate students for helping to keep my thinking on track.

I have benefited from discussions with too many helpful persons to list individually. I am grateful to McMaster University for a good working environment over many years and for post-retirement services and support.

It is assumed that the reader is familiar with phase petrology and modern mineralogy.

1

Introduction

1.1 Defining trace elements

The intent of this book is to examine processes which lead to the accumulation of trace elements in magmatic and metamorphic minerals and rocks, so at the beginning we must consider and define the terms to be encountered.

The first task is to examine what is meant by a *trace element*. In a literal sense it is an element which is present in a rock, mineral or fluid at a low concentration. It is usual in the field of geochemistry to define *major elements* as those which give the sample whatever distinctive character it has, such as its mineralogical make-up; for example, the major elements of cherty limestone would include Ca, C, Si and O. In the case of most common rocks the major elements would include Si, Ti, Al, Fe, Mn, Mg, Ca, Na and K, with abundances in excess of perhaps 1% (in this book % will always be taken as a weight ratio, unless otherwise indicated). Note that O is not usually listed, because the other elements are bound to it.

A number of *minor elements* occur at concentrations usually below 1%; in part they correspond to the presence of accessory minerals such as, apatite (P), zircon (Zr), fluorite (F) etc.

Elements at low concentrations, but which do not give rise to characteristic minerals are classed as *trace elements*. Their manner of occurrence is to be discussed later. The terms used for concentration are %, ppm ($1 \text{ ppm} = 10^{-6} \text{ gg}^{-1} = 0.0001\%$) and gt^{-1} ($1 \text{ gt}^{-1} = 1 \text{ ppm}$).

So far, this is the language of the chemical analyst, and a statement of the kind 'the trace element rubidium comprises 25 ppm of the rock' gives no problem about its intended message: in 1 tonne of the rock there are 25 g of the metal Rb. Sometimes, however, the meaning is not so clear; e.g. to speak of 'the partition of Ni between olivine and clinopyroxene' raises the question of whether Ni, as a metallic element, can somehow occur in the silicate olivine or, in other words, what form Ni must take in order for it to be involved in some sort of reaction or equilibrium between

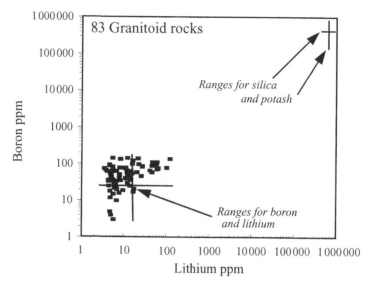

Fig. 1.1 Analyses of 83 granitoid rocks from the Peña Negra complex in central Spain (Pereira and Shaw, 1997) show how trace elements Li and B vary in concentration over two orders of magnitude, whereas the major components SiO$_2$ and K$_2$O vary much less. The trace element concentrations are more sensitive to variations in conditions of origin.

the two minerals; is it present as Ni metal, or as Ni^{2+}, or as NiO, or in some other state? Without further elaboration the statement is not very precise, although useful as a kind of shorthand.

Since the behaviour of trace elements in rocks is the major topic of this book it is desirable to indicate why they merit such attention. In many geochemical studies the approach used is exploratory and, with a particular goal in mind, elemental analyses of rock or sediment samples may serve as variables to test different alternatives and reach conclusions about an origin or a historical evolution. All the elements chosen may, of course, provide useful information, but there are at least three reasons why trace elements are often given special attention in such research:

(i) the lower the concentration of an element, the more likely it is that its behaviour will be regular (*ideal*, in the language of the solution chemist) and not subject to effects linked to its absolute abundance; it may therefore provide information regarding external variables governing the evolution of the rock;

(ii) the range of concentrations is not as restricted or interdependent as major elements (see Fig. 1.1); the latter must sum to 100% and therefore their concentrations are not independent of each other;

(iii) trace elements exhibit a wider range of chemical behaviour as exhibited by their position in the periodic table, compared with the more restricted range of major element chemistry.

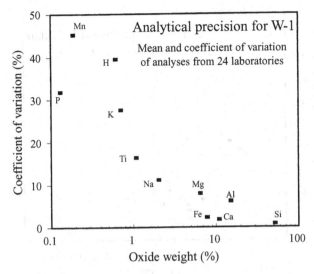

Fig. 1.2 Mean values and coefficient of variation (percentage) for some major and minor elements (expressed as oxides) in the diabase W-1, using data supplied by 24 laboratories (Fairbairn *et al.*, 1951, Table 14).

1.2 The quality of trace element data

Trace elements are inherently at low abundances and consequently they are difficult to analyse with precision and accuracy.[1] Few quantitative data were available before World War II and the quality of analyses has changed greatly since ~1950. In the 1940s a project was initiated in the United States to calibrate two rocks (a granite G-1 and a diabase W-1) by cooperative analyses for major, minor and trace elements from a number of participating laboratories, so that the rocks could serve as standards for precision and accuracy.

The results of the project (see Fairbairn *et al.*, 1951) were disappointing, because the analyses from the participating laboratories showed many discrepancies (for major as well as trace components), leading to a confirmation of the view that rock analysis is a difficult art.

For example, five laboratories respectively reported the Sr concentration in G-1 as 900, 120, 250, 280 and 450 ppm, with similar wide ranges for other trace elements. The variation in such results may be expressed by the *coefficient of variation*, which is the ratio of the standard deviation (*sd*) of the results to their mean value, and is usually expressed as a percentage. For major and minor elements (or oxides) Fig. 1.2

---

[1] It is necessary to bear in mind that *precision* (Ger: *zufälliger Fehler, Reproduzierbarkeit*; Fr: *réproductibilité*) refers to the ability to reproduce an analytical result by multiple measurements using a particular method, whereas *accuracy* (Ger: *Genauigkeit*; Fr: *précision*) is concerned with the ability to measure the 'true' value, without systematic bias dependent on the analytical method in use. The uncertainties introduced from these two causes are sometimes referred to, respectively, as *random error* and *systematic error.*

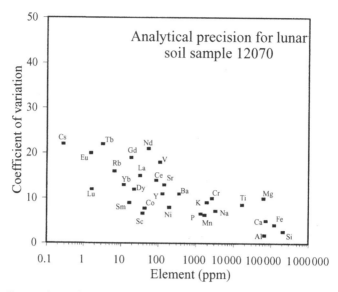

Fig. 1.3 Comparison of analytical error for many elements in the lunar soil sample 12070, as measured in 8–17 laboratories (Morrison, 1971, Table II). The errors are less for Na, Ti, K, Mn, P, than in Fig. 1.2, but remain substantial for elements at the ppm level.

shows that the degree of uncertainty in the diabase W-1 was greater for components at lower concentrations, and this was true also for G-1. Elements plotting in the upper part of Fig. 1.2 can be regarded as having little better than semi-quantitative status.

Strenuous efforts were made to improve the quality of rock analysis in ensuing years. New methods were developed and tested extensively, to verify claims of improved precision and accuracy. By the time that the Apollo missions were returning samples of lunar rocks and soils, better measures of composition could be made. For example, Fig. 1.3 shows that, for a similar group of elements to the ones previously shown, the analytical uncertainty was much less for Mn, P, K and Na, but remained considerable for the trace elements. In the case of the rare earth elements (REE), for which great efforts were invested in obtaining good analyses, Fig. 1.4 shows that the ranges in values obtained, when normalised to chondritic meteorites, were relatively consistent from La to Lu, which is not so evident in Fig. 1.3.

The preceding discussion focuses on precision or reproducibility, but of course Figs. 1.2, 1.3 and 1.4 depict analytical errors which also include systematic or interlaboratory bias; estimates of precision within a single laboratory were, in some cases, markedly superior.

In subsequent years many natural materials have been carefully analysed in many centres, certified with *recommended values* and made available for calibration purposes, as *standard reference materials* (SRM); a journal devoted to these

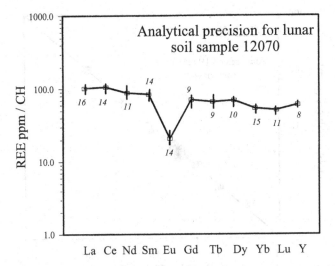

Fig. 1.4 Coefficients of variation of REE analyses (Morrison, 1971, Table II) for lunar soil 12070, from various laboratories, are quite similar in magnitude when the concentrations are normalised to chondritic meteorites (Anders and Grevesse, 1989); numbers of analyses are shown.

matters is the *Geostandards Newsletter*, founded in Nancy, France, in 1977, by K. Govindaraju and published by the Association Scientifique pour la Géologie et ses Applications. Figure 1.5 shows how about 20 such SRMs were used to calibrate the determination of Gd in silicate materials using prompt gamma neutron activation analysis. The linearity of the plot shows satisfactory precision, which by regression analysis was determined (as a coefficient of variation) to be 3–5% of the amount present at concentrations of 1 ppm and greater, and somewhat less at lower abundances.

1.3 Sample heterogeneity

The quality of analytical data depends also on the nature of the samples analysed. The previous discussion has centred on analysis of rocks, yet rocks are mixtures of minerals, and every sample taken will have different mineral proportions. Figure 1.6 sketches a hypothetical volcanic rock composed of large feldspar phenocrysts set in a fine-grained ground-mass, from which samples i, j and k are taken. The first is entirely within feldspar, the second is entirely ground-mass and the third is a mixture of the two; evidently their compositions will be very different. Averaging such compositions will confuse attempts to get a good measure of analytical precision, not to speak of rock composition. This simple example characterises all the problems of sampling inhomogeneous mixtures. The question of how large

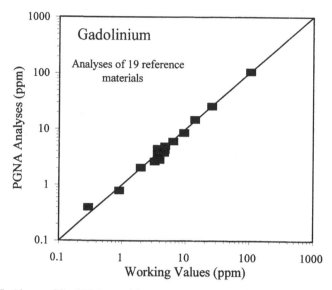

Fig. 1.5 About 20 SRMs, with recommended Gd concentration values (Govindaraju, 1989), used to calibrate a prompt gamma neutron activation analysis method to determine Gd in silicates (Shaw and Smith, 1991). Each of the points represents the average of several PGNA analyses plotted against the recommended value for that SRM. The well-defined linearity shows that precision is satisfactory, except at the lowest concentration where some curvature is apparent.

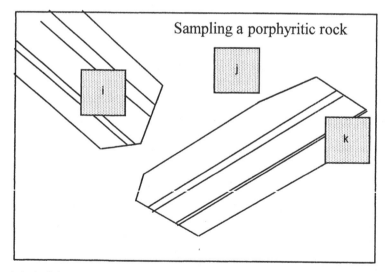

Fig. 1.6 An inhomogeneous volcanic rock composed of large feldspar phenocrysts in a fine-grained ground-mass or matrix. Sample i takes material only from the phenocryst, j is only in matrix while k takes a mixture of both. Analyses of i, j and k will differ greatly for some constituents and such data will confuse attempts to measure analytical precision, as well as attempts to get a good estimate of the rock's composition.

a sample is needed to get a good estimate of rock composition has been treated by many authors, since this is a central problem for economic geologists in estimating ore grade (see for example Gy, 1954), but for geochemical purposes the approach of Laffitte (1957) is convenient.

Thus, a certain volume of rock having weight W consists of various minerals in proportions $p_1, p_2, \ldots, p_j$; a sample of weight w_s is taken and crushed to a fine grain size and then analysed for the element of interest. If

X is the concentration in the whole volume,
x_i is the concentration in grain i, whose weight is w_i,
N is the number of grains in the whole volume,
n is the number of grains in the sample w_s

then the variance associated with the determination of the concentration of the element in the sample is

$$s^2 = \left(\frac{W}{w} - 1\right) \sum \left(\frac{w_i}{W}\right)^2 (x_i - X)^2 \tag{1.1}$$

The summation is over the N grains in the whole volume. If it is assumed that the grains are all about the same size, then the variance reduces to

$$s^2 = \frac{\text{average value of } (x_i - X)^2}{n} \tag{1.2}$$

Since the mineral proportions in W are $p_1, p_2, \ldots, p_j$, and the element concentrations in the minerals are $x_1, x_2, \ldots, x_j$, then

$$s^2 = \frac{p_1(x_1 - X)^2 + p_2(x_2 - X)^2 + \cdots + p_j(x_j - X)^2}{n} \tag{1.3}$$

This expression is now in its most useful form, because it permits estimating how many grains should be taken to achieve a chosen precision level for the element concentration.

For example, suppose that a granodiorite contains 40% plagioclase containing 6% CaO, and 60% other minerals (quartz, mica, orthoclase) containing 0% CaO. Then the rock contains 2.4% CaO, and the sampling variance will be

$$s^2 = \frac{0.4(6.0 - 2.4)^2 + 0.6(0.0 - 2.4)^2}{n} = \frac{9.1}{n} \tag{1.4}$$

Suppose that a sample of 100 grains is chosen, where the grain size is 1 mm^3 with density 2.8 g cm^{-3}; the sample thus weighs 0.28 g. The standard deviations $\approx 0.3\%$ and there will be approximately 95% probability that an analysis of CaO in this sample will lie in the range $2.4 \pm 2 \times 0.3$, or 1.8 to 3.0%. If better precision is required then more grains must be taken, which of course increases the weight of

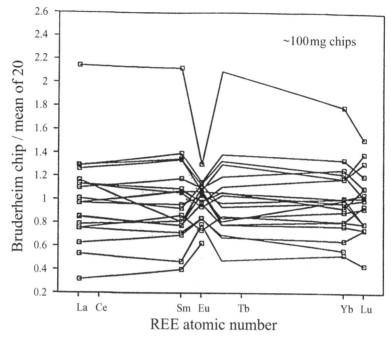

Fig. 1.7 Sampling heterogeneity shown in REE analyses of fragments of the meteorite Bruderheim (Haas and Haskin, 1991). The fragment with the highest REE concentrations is particularly rich in phosphate minerals.

the material to be analysed. For example 2500 grains will reduce the expected range to $2.4 \pm 0.12\%$, but the weight of this sample now $\approx 7\,g$.

A treatment of sampling error by Kleeman (1967) covers similar topics to Laffitte but is more restricted in scope.

An experimental study of sample heterogeneity has been made by Haas and Haskin (1991). Careful analysis of 100 mg fragments of the L6 chondritic meteorite Bruderheim showed striking differences in the REE concentrations as a consequence of mineral variations from one fragment to another (Fig. 1.7). The authors comment as follows:

Note the extensive spread in concentrations, owing mainly to different proportions of phosphate minerals among the different fragments positive and negative Eu anomalies result from different proportions of plagioclase, clinopyroxene and phosphate (Haas and Haskin, p. 16)

Because of rarity and (often) small size, meteorite analyses are often made on samples so small that grain-size heterogeneity must introduce significant error, as this study shows, but the topic is seldom discussed.

One conclusion emerges from this discussion, which is that the geochemist must always bear in mind the heterogeneity of natural materials when sampling is conducted.

1.4 Mineral analysis for trace elements

1.4.1 Sampling

If a mineral specimen is coarse-grained, as in a pegmatite or vein deposit, a specimen for analysis may be taken as a chip sample and subsequently crushed to a fine powder. Care must be taken to avoid contamination of the powder by the trace element of interest from crushing tools or machinery, from any sieves used for sizing the powder, and from residual powder from previous samples; it may be desirable to test for contamination by processing in the same manner some sterile material, such as silica, and subsequently analysing it as a 'blank' to measure the degree of contamination introduced.

More often, the mineral of interest is a rock component and various procedures are available. The oldest approach is to crush the rock to a powder whose fragments are smaller than the mineral grain-size, and then separate the mineral of interest from other components, using heavy liquids, or magnetic separation methods, often followed by hand-picking with the aid of a binocular microscope. The resultant powder may then be analysed for the element of interest.

Such procedures can be very effective, but may also be very tedious and time-consuming. In addition, there are clearly dangers of introducing contamination of various kinds, particularly as incompletely separated grains of other minerals. Until recently, however, only this kind of approach was available, and many of the early measures of element partition coefficients (see p. 23) between the minerals of a rock relied upon analyses of two different mineral powders extracted from a rock.

The inherent assumption of these methods is that a mineral is a clean, homogeneous phase, all of whose grains will behave the same way during separation; in fact, petrographers have always known this to be only an approximation. Mineral grains are commonly zoned in chemical composition, the different zones having different magnetic and density properties and thus behaving differently during separation. Even more importantly, mineral grains usually carry smaller inclusions of different kinds, some too small to be identified by their optical properties. So the chemical composition of a mineral powder really represents the aggregate of the mineral itself plus the various natural contaminants which it has picked up and trapped.[2]

---

[2] The third law of geochemistry states that 'a mineral is not only a structural sieve for elements; it is a dirty sieve' (Shaw, 1964, p. 114).

1.4.2 Chemical analysis

The progress in geochemistry which has been seen since World War II has been very largely dependent on developments in analytical methods. The details are of great interest, but the subject is too vast to review in detail here, and lies outside the scope of this book. Some of the methods employed, however, are introduced in the following paragraphs, together with references where more detailed treatments are to be found.

Early analyses for trace elements employed a variety of methods. In North America many colorimetric methods were developed, relying on the use of a specific reagent to generate a coloured complex in a solution containing a particular element, measuring the complex concentration by its colour absorbance with a spectrophotometer.

In Europe, V. M. Goldschmidt's laboratory in Göttingen and, later, in Norway, pioneered the use of the DC arc to excite characteristic optical emission spectra of all the elements in a sample of powder, isolating the lines of interest from a particular element by prism or grating dispersion and estimating abundances by the image intensity on photographs. Optical emission spectroscopy (OES) became a major analytical tool in the mid twentieth century in many fields and industries, and geochemical applications were developed especially in L. H. Ahrens' group at the Massachusetts Institute of Technology. Optical emission spectroscopy, however, eventually yielded precedence to the more precise and more reproducible X-ray fluorescence (XRF) analysis and neutron activation analysis (NAA), but remains in use in wet laboratories as the modified technique of atomic-absorption analytical spectroscopy (AAAS).

The emergence of truly quantitative geochemical data accompanied the development of microbeam analytical tools. The electron microprobe (EMP), secondary ion mass spectrometry (SIMS) (or ion microprobe (IMP)) and laser ablation inductively coupled plasma mass spectrometry (LA-ICP-MS) permitted analysis of a mineral or a glass matrix in a polished thin section, at low or trace concentration levels, without the necessity of a physical separation.

The development of XRF analysis by R. Castaing in France into an instrument whose electron excitation beam can be focused onto a spot of 10–100 μm diameter created the electron microprobe (EMP) and made it possible to analyse mineral grains in a thin or polished section, and even detect elements present in inclusions visible in thin sections but too small to identify otherwise. For example, the analysis of a separated pyroxene powder from a rhyolite lava (Fig. 1.8) revealed the presence of significant concentrations of the REE, but EMP analysis showed this to be attributable to minute inclusions of an REE-rich mineral. Inferences about the role

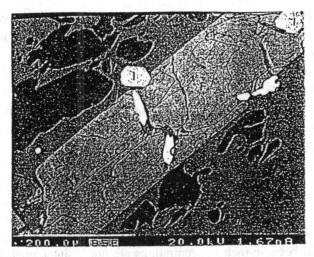

Fig. 1.8 A back-scattered electron image of a high-silica rhyolite from Sierra La Primavera, Mexico. Analyses of the lath-shaped crystal of hedenbergite revealed significant REE concentrations, owing to the presence of small (3–25 μm) inclusions (marked 'C') of the REE-rich mineral chevkinite (Michael, 1988, Fig. 1).

of the REE in pyroxene were thereby compromised. Many such examples have been found, and many earlier interpretations of trace element behaviour (including numerical values of partition coefficients) have had to be modified.

Many examples will be given in the following pages of trace element analyses by EMP and this method initiated a new, quantitative era in geochemistry. Nevertheless, it proved difficult to carry measurement levels for many elements into the ppm domain, for X-ray methods are less sensitive than optical ones; the EMP has remained at its best with analyses for major and minor elements.

The development of the mass spectrometer into a highly sensitive and precise instrument for measuring isotopic ratios during the latter half of the twentieth century facilitated spectacular improvements and discoveries in geochronology. Initially used mostly for the analysis of liquids and gases, adaptation for the direct analysis of solids attached to heating filaments created the solid-state mass spectrometer, which proved to be of great value in earth science. Its marriage with the EMP, but using an ion beam rather than electrons, led to an analytical tool which could measure isotopic abundances in small-beam targets on the surface of a polished specimen i.e. IMP or SIMS. Progess in instrumentation has been slow, but towards the end of the twentieth century a number of laboratories began to produce analyses of acceptable quality. Several good reviews of the exploitation of the IMP in the earth sciences have been given by Nobu Shimizu, the first in 1978 (Shimizu *et al.*), updated in several more recent papers (e.g. Johnton *et al.*, 1990).

Inductively coupled plasma mass spectrometry developed out of optical emission spectroscopy. The sample (liquid or solid) is ionised to a plasma in an arc-like discharge in an inert gas and the ion fractions separated by mass spectrometry. A microbeam instrument was developed by using a focused laser beam to ablate and excite ions directly from a polished specimen (LA-ICP-MS). This method has been in development since the 1980s, but is now capable of accurate and precise analysis at low concentrations. A good account of principles and practice is given by Sylvester (2000).

1.5 Summary

A trace element is a constituent of geological materials which occurs at a low concentration and not in a mineral characteristic of the element. Trace elements are often of interest because their concentrations are not so interdependent as major elements, and may thereby provide useful information.

The quality of trace element analyses was poor until the 1950s; it has been improving since, especially since the introduction of microbeam analytical methods.

Variation detected in repeat trace element analyses in rocks may be determined as much by sampling heterogeneity as by analytical error.

Because of sampling heterogeneity, geochemical study of rocks may best be made from mineral analyses. The fact, however, that completely pure minerals are impossible to obtain, impedes such studies. Instrumental analysis using an excitation beam focused on a small mineral target helps overcome this problem and improves analysis quality.

References

Anders, E. and N. Grevesse (1989) Abundances of the elements: meteoritic and solar. *Geochimica et cosmochimica acta*, **53**, 197–214.

Fairbairn, H. W., W. G. Schlecht, R. E. Stevens, W. H. Dennen, L. H. Ahrens and F. Chayes (1951) A cooperative investigation of precision and accuracy in chemical, spectrochemical and modal analysis of silicate rocks. *United States Geological Survey Bulletin*, **980**.

Govindaraju, K. (1989) Compilation of working values and sample description for 272 geostandards. *Geostandards Newsletter*, **13**, Special Issue, 1–113.

Gy, P. (1954) Erreur commise dans le prélèvement d'un échantillon sur un lot de minérai. *La révue de l'industrie minérale*, Section A, 31–65, April 1954.

Haas, J. R. and L. A. Haskin (1991) Compositional variations among whole-rock fragments of the L6 chondrite Bruderheim. *Meteoritics*, **26**, 13–26.

Johnson, K. T., H. J. B. Dick and N. Shimizu (1990) Melting in the oceanic upper mantle: an ion microprobe study of diopsides in abyssal peridotites. *Journal of Geophysical Research*, **95, No. B3**, 2661–78.

Kleeman, A. W. (1967) Sampling error in the chemical analysis of rocks. *Journal of the Geological Society of Australia*, **14**, 43–7.

Laffitte, P. (1957) *Introduction à l'étude des roches métamorphiques et des gites métallifères*. Paris: Masson et Cie.

Michael, P. J. (1988) Partition coefficients for REE in mafic minerals of high silica rhyolites: the importance of accessory mineral inclusions. *Geochimica et cosmochimica acta*, **52**, 275–82.

Morrison, G. H. (1971) Evaluation of lunar elemental analyses. *Analytical Chemistry*, **43**, 22A–31A.

Pereira, M. D. and D. M. Shaw (1997) Behaviour of boron in generation of the Peña Negra anatectic complex, central Spain. *Lithos*, **40**, 179–88.

Shaw, D. M. (1964) *Interprétation géochimique des éléments en traces dans les roches cristallines*. Paris: Masson et Cie.

Shaw, D. M. and P. L. C. Smith (1991) Concentrations of B, Sm, Gd and H in 24 reference materials. *Geostandards Newsletter*, **15**, 59–66.

Shimizu, N., M. P. Semet and C. J. Allègre (1978) Geochemical applications of quantitative ion-microprobe analysis. *Geochimica et cosmochimica acta*, **42**, 1321–34.

Sylvester, P. (ed.) (2001) Laser-Ablation ICPMS in the Earth Sciences: principles and applications. *Mineralogical Association of Canada Short Course Series*, **29**.

2
Partition coefficients

2.1 Introduction

This chapter will be concerned with the concept of the *partition coefficient* of an element, or the ratio of its concentration between two phases, which is found to be a fruitful variable for studying element distributions. The phases may be minerals, which are not pure chemical compounds but mix-crystals or solid solutions, or may be liquids or gases. Two or more constitute a heterogeneous phase system, and are studied using the principles of thermodynamics. Such systems may be closed, i.e. of constant mass, or open to loss or gain of material. Their study may be restricted to equilibrium conditions or, in certain circumstances, to fractionation processes. Heterogeneous systems may also be subjected to kinetic processes, but this subject has received less attention in geochemistry because of the immensely long time periods often required for geological change.

The idea of the partition coefficient will be developed in the first part of this chapter in its simplest terms, showing that for a trace constituent it may be taken as a constant in certain circumstances. It will frequently be used in this way in subsequent chapters when considering the simplified behaviour of elements during processes of melting and crystallisation. The second part of the chapter will review the causes of variation in trace element partition coefficients, mainly as a result of the influence of other components.

One of the earliest papers devoted to the use of the partition coefficient was by Frederickson (1966) who described it as 'a new tool for studying geological problems' although it had already been explored in some detail by McIntire (1963).

A *partition coefficient* (D) is defined, following Beattie *et al.* (1993) as the ratio of the concentration (c) of a component (i) between two phases (α and β). If a phase of weight W_i contains w_i g of component i, its concentration is $c_i = w_i / W_i$, and the coefficient is defined (see Beattie *et al.*, 1993) by

$$D_i^{\beta-\alpha} = \frac{c_i^{\beta}}{c_i^{\alpha}} \qquad (2.1)$$

where

$$\sum_p c_i^\beta = \sum_p c_i^\alpha = 1 \tag{2.2}$$

for the p components of each phase.

The theory will be considered in two ways – as a system of two solutions with a common solute, or two reacting phases.

2.2 Solutions with a common solute

If there are traces of Rb in coexisting microcline feldspar and albite feldspar, one way to regard the system is to consider that a hypothetical Rb–feldspar occurs in solid solution in each of the two solid phases. This means that the solids may be treated as a ternary system, whose end-members are K–feldspar, Na–feldspar and Rb–feldspar, which may be referred to as K–f, Na–f and Rb–f respectively. An alternative description of the system is a substitution of Rb^+ for K^+ and Na^+, at a cation site in the feldspar structure. If Rb is only one of many elements occurring in similar ways in the two feldspars, then the system is not ternary, and it may be necessary to take the other components into account.

We assume a system at a given pressure (P) and temperature (T), consisting of two phases α and β, each of which is a solution containing dissolved components. For component i the condition for equilibrium partition is that its chemical potential μ be equal in the two phases

$$\mu_i^\beta = \mu_i^\alpha \tag{2.3}$$

If we write a for thermodynamic activity, then this equation can be expanded to

$$\mu_i^{0\beta} + RT \ln a_i^\beta = \mu_i^{0\alpha} + RT \ln a_i^\alpha \tag{2.4}$$

whence

$$\ln \frac{a_i^\beta}{a_i^\alpha} = -\frac{\mu_i^{0\beta} - \mu_i^{0\alpha}}{RT} \tag{2.5}$$

where R is the gas constant, and the term $-(\mu_i^{0\beta} - \mu_i^{0\alpha})$ is the Gibbs free energy of transfer of 1 mole of component i from phase β to α. At a given T and P the right-hand term is constant and may be written as $\ln K_i$, so

$$\frac{a_i^\beta}{a_i^\alpha} = K_i \tag{2.6}$$

This expression says that the ratio of the activities of the solute i in the two phases is a constant at a given P and T. It was first enunciated by D. Berthelot and later developed by W. Nernst.

The activity is related to molar concentration[1] x by an activity coefficient γ so that

$$a_i = x_i \gamma_i \tag{2.7}$$

If i is a major component then γ will vary with concentration. In dilute or low-concentration solutions, i.e. of trace substances, then the γ values closely equal unity, the activity equals the concentration and the system is said to be *ideal*.

Then

$$\frac{x_i^\beta}{x_i^\alpha} = K_{i(P,T)} \tag{2.8}$$

In this case the ratio of the molar concentrations of a trace component in the two phases is constant at a given pressure and temperature and does not vary with the concentration; this is sometimes referred to as the Nernst law. The ratio will here be represented by D_* or, more completely $D_{*i}^{\beta-\alpha}$, and is the *molar partition (or Nernst) coefficient*

$$D_{*i}^{\beta-\alpha} = \frac{x_i^\beta}{x_i^\alpha} \tag{2.9}$$

and

$$\sum_p x_i^\beta = \sum_p x_i^\alpha = 1 \tag{2.10}$$

where there are p components.

It is necessary to look at the relationship between c_i and x_i, i.e. the weight and molar concentrations. In any phase of weight W the weight of component i is w_i and if the formula weight of i is M_i, then the number of gfu of i is n_i, where

$$n_i = \frac{w_i}{M_i} \tag{2.11}$$

and so

$$x_i = \frac{n_i}{\sum_p n_i} = \frac{w_i}{M_i} \bigg/ \sum_p \frac{w_i}{M_i} \tag{2.12}$$

---

[1] It is more appropriate in some cases to refer to the *gram-formula unit* (gfu) concentration, especially where cation substitution is concerned.

But also

$$w_i = c_i W \qquad (2.13)$$

and so

$$x_i = \frac{c_i}{M_i} \bigg/ \sum_p \frac{c_i}{M_i} \qquad (2.14)$$

where the summation is carried out over all the p components i. Comparing Eqs. (2.1), (2.9), (2.14) it is evident that, although the sums of the x_i and of the c_i are each unity, the weight concentrations are empirically more useful, because they do not require any knowledge or supposition regarding molecular or formula units, but are derived directly from analytical data.

It follows from Eq. (2.14) that

$$D_{*i}^{\beta-\alpha} = D_i^{\beta-\alpha} \cdot \frac{\sum\limits_p c_i^\beta \big/ M_i}{\sum\limits_p c_i^\alpha \big/ M_i} \qquad (2.15)$$

The last term on the RHS is a constant for the system, so the Nernst partition coefficient is a constant for a given temperature, pressure and abundance of components other than i.

Returning to the example used at the beginning of this section, it is an artificial distinction to consider Rb–f as a 'solute', when it is the same in kind as K–f and Na–f. If the two feldspars behave as ternary solid solutions, then the molar concentration of Rb–f is x_{Rb-f} and it is clear that

$$x_{Rb-f}^\beta = \frac{n_{Rb-f}^\beta}{n_{K-f}^\beta + n_{Na-f}^\beta + n_{Rb-f}^\beta} = \frac{n_{Rb}^\beta}{n_K^\beta + n_{Na}^\beta + n_{Rb}^\beta} \qquad (2.16)$$

so the molar concentration in any one of the phases present equals the concentration ratio of the numbers of gfu of the substituting cations.

2.3 Reacting solutions: law of mass action

The discussion in the preceding section is not appropriate for considering the partition of, for example, Rb between microcline feldspar and phlogopite biotite, because of the absence of a Rb compound soluble in both minerals. If, however, we assume that Rb–f is present in microcline and that Rb–mica (Rb–m) is present in phlogopite, then the interaction of feldspar and mica solid solutions can be represented by the following exchange

$$(K\text{–f})_{\text{feld}} + (Rb\text{–m})_{\text{mica}} \rightleftharpoons (K\text{–m})_{\text{mica}} + (Rb\text{–f})_{\text{feld}} \quad (2.17)$$

If the terms in parentheses are pure compounds then, at a given P and T, the reactants on the LHS interact, giving rise to the products (RHS), until equilibrium is reached and the activities are related by an exchange constant K, so that

$$\frac{a_K^{\text{mica}} \cdot a_{Rb}^{\text{feld}}}{a_{Rb}^{\text{mica}} \cdot a_K^{\text{feld}}} = K_{Rb-K} \quad (2.18)$$

or

$$\frac{a_{Rb}^{\text{feld}}}{a_{Rb}^{\text{mica}}} = K_{Rb-K} \cdot \frac{a_K^{\text{feld}}}{a_K^{\text{mica}}} \quad (2.19)$$

This equation expresses the law of mass action.[2]

Here now is a difference from Eqs. (2.6) where the term in activities of K-bearing compounds does not appear. Nevertheless, if we now stipulate that Rb is at a low concentration, and that the same is true for any other exchange reactions taking place, then the activity of the K-compound in each phase will be close to unity and so

$$\frac{a_{Rb}^{\text{feld}}}{a_{Rb}^{\text{mica}}} \approx K_{Rb-K} \quad (2.20)$$

Also, activity will be approximately equal to molar concentration for a trace component, so

$$\frac{a_{Rb}^{\text{feld}}}{a_{Rb}^{\text{mica}}} \approx \frac{x_{Rb}^{\text{feld}}}{x_{Rb}^{\text{mica}}} = D_{*Rb}^{\text{feld–mica}} \approx K_{Rb-K} \quad (2.21)$$

This equation may be compared with Eqs. (2.6) and (2.9). The relationship between the partition coefficient and the molar coefficient in Eq. (2.21) is again similar to that expressed in Eq. (2.15), but which uses normalising formula units M_i and M_j for feldspar and mica respectively.

Figure 2.1 shows an example of a series of analyses of V in mineral pairs from rocks of the same metamorphic grade. The slope of the linear trend defines a partition coefficient which is nearly constant for concentrations which, although at trace levels, range over an order of magnitude, and which therefore conforms to the pattern of Eq. (2.21).

The approach used in the preceding paragraphs is general, and may be applied to reacting solutions of any kind – both solid and liquid. For example, the substitution

---

[2] In many publications on mineral equilibria the exchange coefficient is called K_D, and is referred to as the distribution coefficient, but this is obsolete (see Beattie *et al.*, 1993).

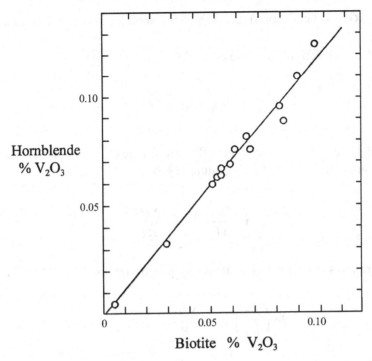

Fig. 2.1 Distribution of V concentrations, expressed as V_2O_3, between coexisting biotite and hornblende pairs, from 14 rocks of the same metamorphic grade (Kretz, 1959, Fig. 7). The linear trend shows that the partition coefficient is essentially constant.

of Ni^{2+} for Mg^{2+} in a solid (s), such as olivine, crystallising in equilibrium from a magma (m), may be represented by an equation similar to (2.21)

$$\frac{x_{Ni}^s}{x_{Ni}^m} = D_{*Ni}^{s-m} \approx K_{Ni-Mg} \tag{2.22}$$

Similar equations may be formulated for metasomatic reactions, e.g. where a fluid carrying a trace element, such as Zn, permeates and replaces a limestone to form sphalerite.

2.4 Effects of variation of temperature and pressure

Applying the identity that $\mu_i = \mu_i^0 + RT \ln a_i$ to Eq. (2.19) leads to the following expression

$$\frac{a_{Rb}^{feld} \cdot a_K^{mica}}{a_{Rb}^{mica} \cdot a_K^{feld}} = e^{-\frac{\left(\mu_K^{0\ mica} + \mu_{Rb}^{0\ feld} - \mu_K^{0\ feld} - \mu_{Rb}^{0\ mica}\right)}{RT}} = e^{-\frac{\Delta G^0}{RT}} \tag{2.23}$$

where $-\Delta G^0$ is the Gibbs free energy of complete conversion of reactants to products.

Thus the mass action exchange coefficient can be written

$$\ln K = -\frac{\Delta G^0}{RT} \tag{2.24}$$

and this may be used to estimate the effect of change of T or P on the reaction.

For example, the effect of temperature is given by the following

$$\left(\frac{\partial \ln K}{\partial T}\right)_P = \frac{\Delta H^0}{RT} \tag{2.25}$$

and integration shows that ln K is inversely proportional to T

$$\ln\left(\frac{K_2}{K_1}\right) = -\frac{\Delta H^0}{R}\left(\frac{1}{T_2} - \frac{1}{T_1}\right) \tag{2.26}$$

These two equations show that exchange coefficients, and partition coefficients in systems which resemble Eq. (2.21), will vary with T and thus might be used as geothermometers. Many attempts have been made to devise geothermometers of this kind, to help solve petrological problems. Perhaps the first to attempt a quantitative treatment was T. F. W. Barth (1956), who tried to establish a geothermometer based on alkali partition between albite and orthoclase.

More successful was Eugster (1955), who showed a clear (and nearly linear) correlation of $Cs^+ - K^+$ exchange with temperature, between sanidine and aqueous solution (Fig. 2.2). This work was extended (Beswick, 1973) to the exchange between phlogopite mica, sanidine and aqueous solution, giving the coefficient $D_{Cs}^{phl-fluid}$ as a function of temperature. Taking the ratio with $D_{Cs}^{san-fluid}$ gives the coefficient $D_{Cs}^{phl-san}$, which may be applied to natural assemblages of mica and potash feldspar as a geothermometer. Of course, the experimental system was chemically much simpler than natural minerals, and the application required assumptions of equilibrium. The results were not encouraging but the principle was developed with greater success by Wones and Eugster (1965).

An example, however, from a multi-component system is seen in Fig. 2.3, where experimental values of the olivine–melt equilibrium exchange of six elements with Mg show clear dependence on the reciprocal of temperature, in spite of experimental scatter. The effects of other components on D-values will be further examined below.

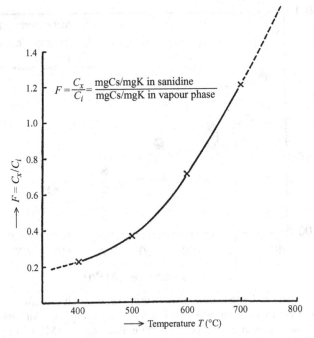

Fig. 2.2 Variation of the equilibrium constant K (here shown as F) for the exchange of Cs for K between sanidine and a fluid solution, as a function of temperature (Eugster, 1955, Fig. 3).

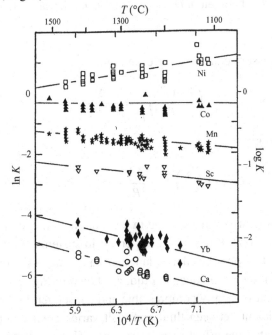

Fig. 2.3 Plot of $\ln K$ against 10^4/temperature for experimental equilibration of olivine, doped with various elements at 0.5–4% concentrations, with basaltic melt at 1 bar pressure. The equilibrium constant K is for exchange of the added element with Mg^{2+} (Colson *et al.*, 1988, Fig. 8).

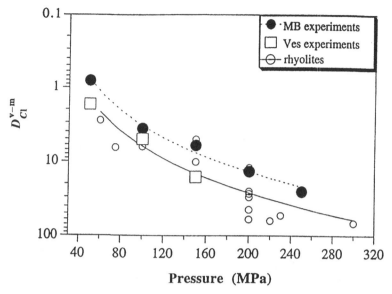

Pressure (MPa)

Fig. 2.4 Experimental measurements of the vapour–melt partition of chlorine at different pressures (Signorelli and Carroll, 2000, Fig. 6). An NaCl-bearing aqueous solution was equilibrated with phonolitic melts, (MB) from Montaña Blanca, Tenerife and (Ves) from Vesuvius, Italy, at temperatures of 860–890 °C. The rhyolites shown by circles were taken from cited papers. It is clear that increasing pressure will increase the Cl-content of vapour relative to the melt, but the relationship between $\ln D$ and pressure is not linear (note the reciprocal scale in D).

The effect of pressure change is governed by

$$\left(\frac{\partial \ln K}{\partial P} \right)_T = -\frac{\Delta V^0}{RT} \tag{2.27}$$

which integrates to give

$$\ln \left(\frac{K_2}{K_1} \right)_T = \frac{\Delta V^0}{RT}(P_2 - P_1) \tag{2.28}$$

where ΔV^0 is the molar volume change in the mass action equation, showing that $\ln K$ is proportional to P. It is seldom important in systems containing only solid phases to consider this pressure dependance, because the ΔV^0 term is small. But when one of the phases is an aqueous fluid, as shown in Fig. 2.4 (from Signorelli and Carroll, 2000), it is readily compressible and consequently ΔV^0 is larger, and the partition coefficient between fluid and melt varies greatly with pressure. But it should be noted that the effect of pressure on $\ln K$ is not linear in the example shown in Fig. 2.4.

The near-constancy of *D*-values expected from Eqs. (2.9) or (2.21) applies to cases where additional components, if present, do not show variation from sample to sample. In systems where such components must be treated as variables complications are to be expected and will be examined later in the chapter.

2.5 Measurement of partition coefficients

Subsequent chapters will make frequent use of *D*-values and, although their measurement will not be discussed in detail, this section presents a brief overview of a rapidly evolving subject.

It is not easy to measure partition coefficients for systems relevant to igneous and metamorphic geology and virtually nothing was done before 1950. The early measurements were made using natural materials – minerals and rock matrices, because of the difficulty of making experimental measurements in the laboratory.

The first useful measurements of partition coefficients were made using volcanic rocks, by analysing phenocrysts and their surrounding matrix. This required the assumption that the two formed in equilibrium, and remained so subsequently; the evidence was seldom strong. The analyses were carried out on powders extracted from crushed rock after meticulous separation routines by highly skilled individuals. The measurements were beset with possible errors at every step, but for many years little else was available. It is not surprising that sceptical comments were expressed.[3] The sources of error were principally questions of:

(i) purity;
(ii) homogeneity;
(iii) equilibration;
(iv) analysis error.

With regard to purity, it was seen in Chapter 1 that minute inclusions of an accessory mineral might be present in the mineral of interest; the example of the REE-bearing accessory chevkinite was mentioned as a contaminant of major minerals, which could seriously disturb measurements of partition. Cameron and Cameron (1986) show that the accessory allanite[4] can seriously affect measurements of coefficients of the REE for hornblende, including the introduction of a prominent Eu anomaly (Fig. 2.5). Considering the role of a mineral such as hornblende during melting, however, the relevant variable will be the distorted coefficient of hornblende-plus-inclusions, rather than that of pure hornblende (see Bea, 1996).

---

[3] 'At the present time, the only way to get useful partition coefficients is to criticize all existing results and make a choice between contradictory data' (Allègre *et al.*, 1977, 66).
[4] Chevkinite is a variety of allanite.

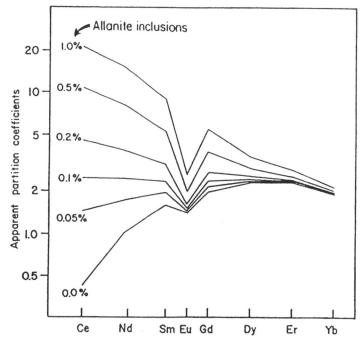

Fig. 2.5 Measurements of the partition coefficient hornblende–volcanic glass for the REE can be distorted by even small inclusions of allanite. In this example the allanite even brings an Eu anomaly, virtually absent from the hornblende (Cameron and Cameron, 1986, Fig. 4).

Homogeneity is a concern because, if the composition of a mineral varies from core to rim, then the composition of a crushed sample may not quantitatively reflect that variation. In addition, such variation would indicate that the third concern, i.e. equilibration, has not been achieved.

A development which helped to improve the analytical quality of such measurements was the availability of microbeam analytical tools. The EMP, SIMS (or IMP) and LA-ICP-MS permit analysis of a mineral or a glass matrix in a polished thin section, without the necessity of a physical separation, at low or trace concentration levels. For experimental measurements of partition, a second development was the improvement in techniques for experimental syntheses at high T and P; the microbeam techniques have proved invaluable in this field also.

For example, Figure 2.6 shows a comparison between earlier and more recent plagioclase–glass D-values measured by Bindeman *et al.* (1998) for three elements. The new measurements (black squares), made by ion microprobe, are plotted against the anorthite content of the feldspar, and show near-linear correlation. By contrast, the crosses are earlier phenocryst/matrix measurements made on volcanic rocks,

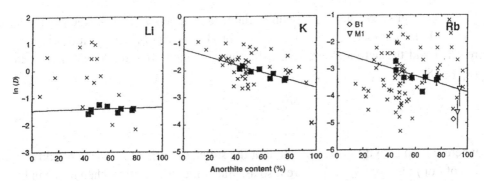

Fig. 2.6 *D*-values for Li, K and Rb, between plagioclase feldspar and coexisting basaltic glass, plotted against the anorthite content of the feldspar (Bindeman *et al.*, 1998). The black squares are new ion probe measurements and show nearly linear trends, without major dispersion. The crosses are older values obtained by various authors by analysing plagioclase phenocrysts and coexisting matrix in volcanic rocks; their dispersion is too large to yield useful values for modelling. Open symbols are electron probe measurements.

and show wide variations – more than three orders of magnitude in the case of Rb; such values clearly have no use in quantitative modelling.

A notable growth in research resulted from improvements in analytical procedures. Progress was recorded in the Sedona conference proceedings (Drake and Holloway, 1978) and also in a special issue of *Chemical Geology* (Foley and Van der Laan, 1994) and a testimonial volume for B.B. Jensen (Austrheim and Griffin, 2000). These works provide good information for readers wanting further details.

2.6 Extended theory

First we will consider the theoretical basis for considering that the equilibrium partition of two elements i and j between two phases must be affected by other components. The reaction can be written as follows

$$pA(i)_\alpha + qB(j)_\beta = pA(j)_\alpha + qB(i)_\beta \qquad (2.29)$$

where α and β are the two phases, A and B are the two compounds, each of which is a solid solution series between the end-members i and j, and p and q are the numbers of stoichiometric units. For example, the phase *feld* (feldspar) is a solid solution of K–f and Rb–f, and the phase *mica* is similarly made up of the compounds K–m and Rb–m, and the equilibration reaction we saw in Section 2.3 is Eq. (2.17)

$$(K-f)_{feld} + (Rb-m)_{mica} \rightleftharpoons (K-m)_{mica} + (Rb-f)_{feld}$$

If the reactants are pure compounds then, at a given P and T, the reactants on the LHS interact, giving rise to the products (RHS), until equilibrium is reached and the activities are related by an exchange constant K, so that, as in Eq. (2.18)

$$\frac{a_{j,\alpha}^{p} \cdot a_{i,\beta}^{q}}{a_{i,\alpha}^{p} \cdot a_{j,\beta}^{q}} = K_{i-j} \tag{2.30}$$

where a is an activity. If we assume that the stoichiometric units each contain *one* ion of i or j then $p = q = 1$; if we also assume ionic dissociation, then $a_{j,\alpha}$ can be replaced by $x_{j,\alpha}\gamma_{j,\alpha}$, where γ is an activity coefficient and $x_{j,\alpha}$ is an ionic fraction, and Eq. (2.30) becomes

$$\frac{x_{j,\alpha} \cdot x_{i,\beta}}{x_{i,\alpha} \cdot x_{j,\beta}} = K_{i-j} \cdot \frac{\gamma_{i,\alpha} \cdot \gamma_{j,\beta}}{\gamma_{j,\alpha} \cdot \gamma_{i,\beta}} \tag{2.31}$$

We can now substitute $x_{i,\alpha} + x_{j,\alpha} = 1$ and $x_{i,\beta} + x_{j,\beta} = 1$ and combining terms on the RHS into c_{i-j}, we get

$$\frac{x_{i,\alpha}}{x_{i,\beta}} = \frac{1 + x_{i,\alpha}(c_{i-j} - 1)}{c_{i-j}} = D_i^{\alpha-\beta} \tag{2.32}$$

Noting that $\frac{1}{c_{i-j}} = c_{j-i}$ this can be written as follows

$$D_i^{\alpha-\beta} = 1 - x_{j,\alpha}(1 - c_{i-j}) \tag{2.33}$$

or, the coefficient for one element is a linear function of the concentration of the other element in one of the phases.

Now let us consider a three-component system, where the cations 1, 2 and 3 substitute for each other in the same two phases. Then we have

$$D_1^{\alpha-\beta} = 1 - x_2^{\alpha}(1 - c_{2-1})$$
$$= 1 - x_3^{\alpha}(1 - c_{3-1}) \tag{2.34}$$

Also we have

$$x_1^{\alpha} = 1 - x_2^{\alpha} - x_3^{\alpha}; \quad x_1^{\beta} = 1 - x_2^{\beta} - x_3^{\beta} \tag{2.35}$$

and the equations can be combined to give

$$D_1^{\alpha-\beta} = 1 + x_2^{\alpha}(c_{2-1} - 1) + x_3^{\alpha}(c_{3-1} - 1) \tag{2.36}$$

and this can be generalised for any number n of substituting cations as

$$D_1^{\alpha-\beta} = 1 + \sum_{i=2,\ldots,n} x_i^{\alpha}(c_{i-1} - 1) \tag{2.37}$$

Table 2.1 *Estimations of $D_{Ni}^{oli-liq}$ in various systems*

D Values	Source	Date	Comments
13–17	Häkli & Wright	1967	Hawaiian basalts
1.46–3.3	Leeman	1973	variation with T °C
4–24	Hart *et al.*	1979	variation with Si, Mg
1.3–1.7	Irvine & Kushiro	1976	experiments at 15 kbar
5–25	Mysen & Kushiro	1979	experiments at 1 bar–20 kbar
2.9–13.6	Takahashi	1978	variation with K_2O, Si
10.35	Nabelek	1980	mean value
1.7–11	Cawthorn & McIver	1977	variation with Mg in komatiites

This equation thus shows that the partition coefficient for an element which can substitute in two phases is a linear function of the concentrations of all the other cations present which can substitute in the same way; actually, the activity terms in the c_{i-1} terms complicate the relationship. If one of the phases is a mineral with two sites, M-1 and M-2, and the element enters the M-1 site, then the summation in Eq. (2.37) extends to all the M-1 site elements, but has no direct application to M-2 site elements; it may well be, however, that the elements occurring at M-2 affect the activity coefficients of element 1, embedded in the c_{i-1} terms.

Application of Eq. (2.37) is difficult, because of lack of knowledge of the mass-action equilibrium constants K_{i-j}, but the equation nevertheless carries within it the potentiality for calculating the influence of temperature and pressure as well as other components.

2.7 Major element effects

It was realised from the beginning of interest in partition coefficients that focus on one particular element or compound would always require attention to other components. Among others, particular attention has been paid to magma properties including major elements, redox properties, volatile content and melt structure.

2.7.1 Olivine–melt partition

Because of their importance in mantle melting and basalt crystallisation, major element effects on the partition of trace elements between olivine and melt have been extensively studied. Table 2.1 shows how diverse the measurements of $D_{Ni}^{oliv-melt}$ have been in recent years, although in many cases the analytical errors were small.

One approach to such variations has been given by Takahashi and Irvine (1981), as follows. Metallic elements substitute into an olivine of general formula Y_2ZO_4 in the Y or the Z site. Of all the cations, 2/3 or 0.667 go into the Y site (octahedral), substituting for Mg, and the authors define this proportion as ψ, the *stoichiometric constant*. Then if the summation is for the components i substituting at the site and if their fractions (g-equivalent, g-formula-unit, mole fraction) are written n_i^s, then it follows that

$$\psi = \sum_i n_i^s \tag{2.38}$$

where, for example, Fe, Ni, Mn, etc. substitute for Mg in olivine. If two of these substituting elements are in equilibrium then

$$K_{i-j} = \frac{D_i}{D_j} = \frac{n_i^s}{n_i^l} \cdot \frac{n_j^l}{n_j^s} = \frac{1}{K_{j-i}} \tag{2.39}$$

so

$$n_j^s = D_i n_j^l K_{j-i} \tag{2.40}$$

$$\psi = n_1^s + \sum_j n_j^s \quad \text{where} \quad j \neq i \tag{2.41}$$

and

$$= D_i n_i^l + D_i \sum K_{j-i} n_j^l \tag{2.42}$$

whence

$$D_i = \frac{\psi}{n_i^l + \sum K_{j-i} n_i^l} \tag{2.43}$$

This equation[5] says that if one knows the structural formula of the solid, the liquid concentrations of element i and the K_D values, then D_i can be calculated. The authors note that the K_D values for olivine–basic melt are fairly well known, e.g.

FeO/MgO   0.28–0.33
MnO/MgO 0.25–0.30
CoO/MgO 0.60–0.80
NiO/MgO   1.0–3.0

So for a particular melt composition one can first calculate the Mg–Fe exchange, taking $K_{FeO-MgO}$ as ≈ 0.3, so that

$$D_{Mg} = \frac{0.667}{n_{MgO}^l + 0.3 n_{FeO}^l} \tag{2.44}$$

[5] Which is a rephrasing of Eq. (2.31).

and then for an element such as Ni, one can calculate

$$D_{Ni} = D_{Mg} K_{NiO-MgO} \tag{2.45}$$

this procedure was tested for REE partition but with only limited success.

Jones (1984) re-examined the olivine–melt partition for Mg, Fe, Mn and Ni, using a statistical approach which allowed for uncertainties in composition; he showed again that trace element D-values can be predicted from the composition of the melt, with considerable success.

A more recent study by Kohn and Schofield (1994) shows that D_{Mn} and D_{Zn} for olivine–melt partition depend in fact on the K_D values, as just described, but these are also variable and depend on the melt structure to a greater extent than was realised in previous studies. This topic will be examined shortly.

Yet another approach to the topic of melt composition effects on trace element partition has been delineated by Colson *et al.* (1988). This applies to systems where good quality D-values are available, as in the case of partition between olivine or pyroxene and basaltic melt. The authors use an exchange such as

$$Mg^{2+}_{oliv} + tr^{2+}_{melt} \rightleftharpoons Mg^{2+}_{melt} + tr^{2+}_{oliv} \tag{2.46}$$

and make use of the fact that the equilibrium constant K_D is made up of two terms – $\Delta H/RT$ and $\Delta S/R$, which can be expressed in terms of cation radius, charge and energy. Using their experimental data for partition of nine elements at various temperatures and constant pressure they carry out a multiple regression analysis which allows them to estimate the K_D values as functions of reciprocal temperature, and from that the values of D_i for both olivine– and pyroxene–melt. A plot of measured vs. predicted coefficients shows good correlation and permits the prediction of coefficients for other elements, not included in the regression. Some of their results are shown in Fig. 2.3.

2.7.2 Redox effects

The effect of varying redox properties is best seen from the partial pressure of oxygen, which has been shown to have an important effect on D-values in systems with a volatile phase, especially for elements with variable oxidation state. One of the most studied of these is Eu, and V. M. Goldschmidt before World War II recognised that this element's behaviour is different in oxidising and reducing conditions, chiefly because the radius of $Eu^{2+} \approx 1.33$ Å whereas that of $Eu^{3+} \approx$ 1.15 Å; when divalent, Eu follows the similarly sized Sr, whereas when oxidised and trivalent it follows the adjacent REE. Drake and Weill (1975) showed this by measuring $D^{plag-melt}$ for the REE under different values of fO_2, the Eu anomaly

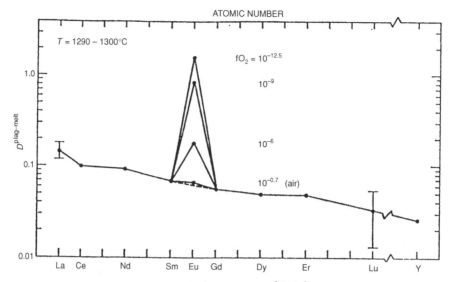

Fig. 2.7 Experiments show that the coefficient $D_{Eu}^{plag-melt}$ is a function of fO_2 and therefore the behaviour of Eu during melting or crystallisation will depend on the oxidation–reduction conditions (Drake and Weill, 1975, Fig. 6).

increasing as fO_2 decreases (Fig. 2.7). In a similar study, Sun *et al.* (1974), for example, measured partition between mineral and basaltic melt for several elements, with fO_2 varying from 10^{-8} to 10^{-14} atm and variable temperatures. Finding linear variation they were able to express their results in the following equations

$$\log D_{Eu}^{plag-melt} = \frac{2460}{T} - 0.15 \log fO_2 - 3.87 \tag{2.47}$$

$$\log D_{Sr}^{plag-melt} = \frac{6570}{T} - 4.3 \tag{2.48}$$

Europium may have charge +2 or +3 whereas Sr cannot be oxidised, so is constrained only by temperature variation. Grutzeck *et al.* (1974) have equations of similar structure for the Eu partition $D^{cpx-melt}$. The significance of an Eu anomaly will be discussed shortly.

A more extreme example is seen in the effect of fO_2 on the partition of siderophile elements between liquid metal and silicate melt (Schmitt *et al.*, 1989). In experiments at 1300 °C the *D*-value for tungsten, as a function of oxygen fugacity, (Fig. 2.8) varies over four orders of magnitude, and changes from a strong preference for the metallic phase at very low fO_2 to a preference for silicate as fO_2 increases.

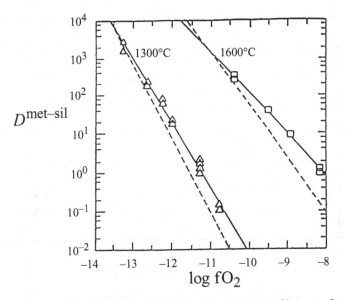

Fig. 2.8 Experiments at 1 bar show that the partition coefficient of tungsten between coexisting metal and silicate melt is controlled by the partial pressure of oxygen, and by temperature (after Schmitt *et al.*, 1989). The relationship is linear, for a given temperature, on a ln–ln plot. The dotted lines give the calculated fO_2 for these experiments.

2.7.3 *Volatile fluids*

A number of studies have been made of systems where a trace element is partitioned between a solid, a melt and a volatile fluid. When the fluid is water, then cations usually show higher concentrations in the fluid than the melt, i.e. $D_i^{fl-m} > 1$, as shown for Li in the measurements by Webster *et al.* (1989) in Fig. 2.9. When other fluid components are present, such as Cl, a decrease in the mole fraction of water ($X_{H_2O}^{fl}$) favours accumulation of the cation in the fluid, with less dependence on temperature. Similar effects of Cl on various cations were reported by others, including Keppler (1996), and also for F (Keppler, 1993). It is likely that the process involves complex formation. Flynn and Burnham (1978) found the partition coefficient of Ce between pegmatite melt and aqueous fluid to be proportional to the cube of the molar Cl concentration in the fluid, as did Reed (1995) for other REE.

We shall look more closely at the effects of a metal forming a complex in the melt. Complex formation has been proposed in the past for many high-field-strength (HFS) elements, particularly in polymerised, silicic melts. Treuil *et al.* (1979) give evidence that this can occur in the case of U. Adsorption spectra show the existence of U in the form of UO_2^{2+} and of U^{4+} in a phosphate complex in basaltic melt.[6] The

---

[6] The captions to Figs. 2 and 3 in Treuil *et al.* (1979) appear to have been interchanged.

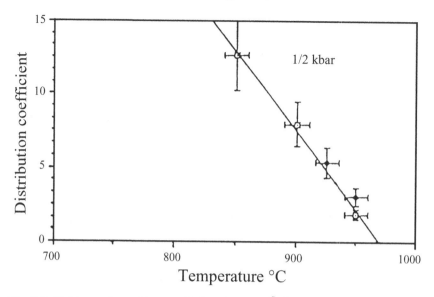

Fig. 2.9 Fluid – melt partition coefficient for Li ($D_{\text{Li}}^{\text{fl}-\text{m}}$) as a function of run temperature, showing strong temperature dependence; the fluid was water-saturated. Experiments where CO_2 was present showed higher Li concentration in the fluid and less dependence on temperature (Webster *et al.*, 1989, Fig. 1).

authors show that the concentration of the cation M of charge $m+$ in the solid is governed by a modified D-value, because the available concentration in the liquid is determined by the stability constant β of the complex, whose formation can be represented by reaction with p units of an anion X of charge $a-$:

$$M^{m+} + pX^{a-} = MX_n^{(m-pa)+} \qquad (2.49)$$

and the customary coefficient D must be modified to D^*. For example, the ion U^{6+} may form the following complex

$$U^{6+} + 2O^{2-} = UO_2^{2+} \qquad (2.50)$$

with stability constant

$$\beta = \frac{\left[UO_2^{2+}\right]}{\left[U^{6+}\right]\left[O^{2-}\right]^2} \qquad (2.51)$$

where the square brackets indicate activities. The true partition coefficient is defined by

$$D = \frac{n_{U^{6+}}^{\text{s}}}{n_{U^{6+}}^{\text{l}}} \qquad (2.52)$$

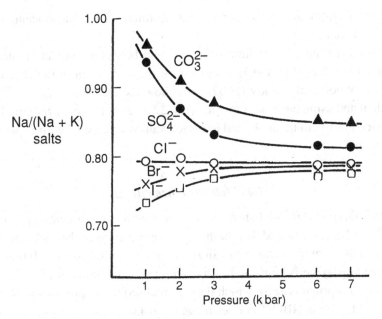

Fig. 2.10 The partition of Na between aqueous solution and alkali feldspar depends on the anion present. Iiyama shows (1970, Fig. 1) the ratio of Na salt to total (Na + K) salts of various anions in hydrothermal solutions which have equilibrated with perthitic feldspar. At low confining pressure Na is strongly enriched (relative to feldspar) in carbonate solutions, relative to chloride solutions. The effects are greatest at low pressure, when the salts used are not completely dissociated. As pressure increases so does the dissociation and the effect on Na partition decreases.

but the modified coefficient D^* uses the total U concentration in the liquid, and so is

$$D^* = \frac{n^s_{U^{6+}}}{n^l_{U^{6+}} + n^l_{UO_2^{2+}}}$$

and consequently, assuming ideality

$$D^* = \frac{D}{1 + \beta \left(n^l_{O^{2-}}\right)^2} \tag{2.53}$$

which means that $D^* < D$, or *complex formation makes the element more incompatible.*

Another effect of anions on partition has been established by Iiyama in a number of articles (e.g. 1970). Experiments on the partition of Na and K between aqueous solutions and alkali feldspar at varying pressures (Fig. 2.10) show enhanced Na in carbonate solutions, relative to chloride solutions, which is attributed to a low degree of dissociation. Increasing pressure tends to promote dissociation and thereby nullify the discrimination between anions. Unfortunately the Na concentrations in the feldspars are not given.

It is also clear that the partition of an element between a silicate mineral and silicate melt is affected by the H_2O content of the melt, even in the absence of a fluid phase. Wood and Blundy (2002) show how the coefficient $D^{cpx-melt}$ for an REE, calculated using the Brice equation (Eq. 2.55 below), may be modified for the water concentration in the melt, and that such values agree well with experimental measurements.

2.7.4 Melt structure effects

Mysen and Virgo (1980) published a study showing a strong linear dependence of $D_i^{oliv-melt}$ for Ni, Ce, Sm and Tm on the bulk composition of the melt and, in particular, with the degree to which the SiO_4 tetrahedra are linked by polymerisation. The degree of polymerisation can be characterised by the number of *non-bridging oxygens* (i.e. un-polymerised), which they normalised to the proportion of tetrahedral cations, to obtain NBO/T. Values range from 4.0 for an unpolymerised simple melt to ∼1.0 for a 3-D network, but the presence of Al and minor oxides such as CO_2, H_2O, TiO_2, P_2O_5 and Fe_2O_3 must be taken into account as network formers, and most values are less than 1.0. Figure 2.11 again shows the dependence of D_{Ni} on NBO/T; the dependence is not matched well by the linear regression line.

Another approach to adjusting measured D-values for melt composition focuses on the cation composition. Following earlier work by Y. Bottinga and D. F. Weill, Nielsen proposed (1985, 1988) a *two-lattice melt model* in which the melt components are *network formers* including Si, Al, Na and K, and *network modifiers*, including Ca, Mg, Fe. Following calculation of the ordinary wt. D values, these could be adjusted by multiplying by the normalised sum of the network modifiers to give a 'compensated' coefficient d^*, constrained only (in theory) by temperature and pressure, as in Fig. 2.12. It is clear that the compensated coefficients have a better linear relationship with inverse temperature, as expected from Eq. (2.26).

Good as was this correction procedure, there has been some controversy about how universal it could be. Ellison and Hess (1990) claimed that the division into formers and modifiers is not clear-cut, and non-ideality within one of the sub-lattices restricts the results from one system being applied in another.

2.8 Influence of the host solid

2.8.1 Substitution site deformation

It has long been known that there is a relationship between the partition coefficient of a metal between two phases, and its cationic size. Masuda (1965) was one of the

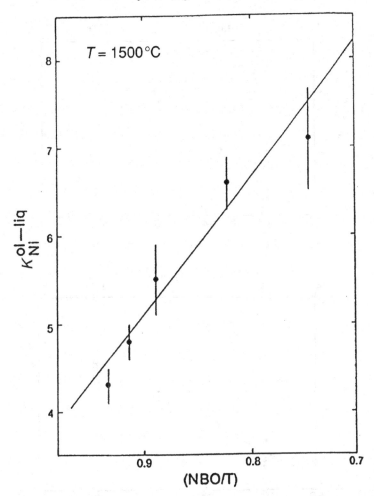

Fig. 2.11 Five isothermal (1500 °C) measurements of the partition coefficient (here called *K*) for Ni, between olivine and melt, recalculated to show the relationship with the ratio NBO/T of non-bridging oxygen to tetrahedral cations (Mysen and Virgo, 1980, Fig. 5). The regression line through the points is said to have an r^2 value of 0.96, but this seems improbable.

first to record this, for elements partitioned between silicate and melt. A number of others (Nagasawa, 1966; Onuma *et al.*, 1968; Higuchi and Nagasawa, 1969; Jensen, 1973) made similar observations in succeeding years. The relationship becomes clear if the coefficients D_i^{s-m}, between solid phase s and melt m for a series of cations of the same valency are plotted in a diagram against their radii r_i. Diagrams of this kind have been referred to as Onuma diagrams; examples are shown in Figs. 2.13 and 2.14. One of the best-known contributions is the last-named reference.[7] The main conclusions of these studies were as follows:

---

[7] Dr Jensen's contributions were celebrated in a dedicated issue of *Lithos* (see Austrheim and Griffin, 2000).

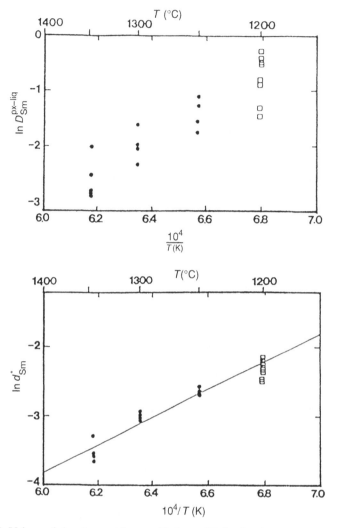

Fig. 2.12 Values of simple partition coefficients (D) for Sm between pyroxene and melt for two sets of data from the literature (top), and coefficients (d^*) corrected by the Nielsen two-lattice model of melt structure; the corrected values show much tighter dependence on inverse temperature (Nielsen, 1988, Figs. 1 and 2) and thus are 'better' values.

(i) D_i^{s-m} depends on the radius and charge on the ion i, and also on the structure of the mineral;

(ii) each mineral has one or more optimum D-values, corresponding to the radius of each structure site; two are marked in Fig. 2.13;

(iii) away from a maximum the D_i^{s-m} vs. r_i relationship approaches linearity;

(iv) $\log D_i^{s-m}$ varies with the square of the radius difference between the ion and the 'size' of the site (see Fig. 2.14);

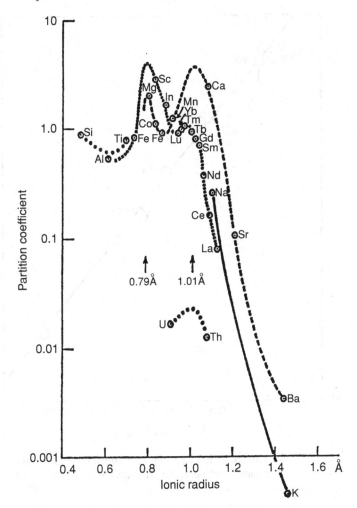

Fig. 2.13 Partition coefficients between augite and basalt matrix (from Onuma *et al.*, 1968) vs. ionic radius (Jensen, 1973, Fig. 1). The estimated maxima at 0.79 Å and 1.01 Å correspond to sites M1 and M2.

(v) major elements do not necessarily plot at the peak of a curve;
(vi) curves of this kind can be used to estimate an unmeasured D_i^{s-m} for an element, knowing its ionic radius and valency.

A useful overview of these and other features was given by Philpotts (1978).

The quantitative nature of the D_i^{s-m} vs. r_i relationship was studied in all the works already cited, but the most fruitful early contribution came from Nagasawa (1966), who reasoned that the substitution of a trace for a major ion at a structural site could be modelled as the elastic energy required to force a sphere of radius $r_0(1 + e)$ into a site of radius r_0, in a solid whose Poisson ratio and compressibility

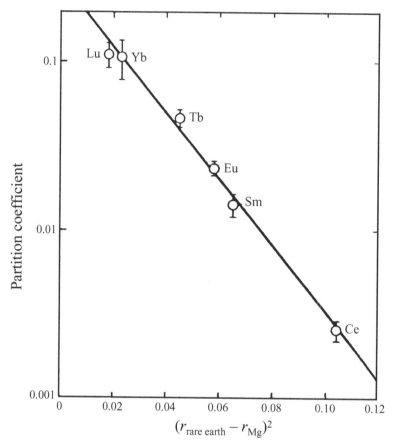

Fig. 2.14 Log D_i^{s-m} for REE partition between bronzite and basalt vs. the square of the difference in ionic radius between the REE and Mg (Onuma *et al.*, 1968, Fig. 3).

are s and K. Then if the partition coefficient between solid and liquid of the trace component is D he found that

$$\ln D = -\frac{6\pi r_0^3(y-1)}{K\,ykT}e^2\left(1+\frac{e}{y}\right) \qquad (2.54)$$

where $y = 3\frac{(1-s)}{(1+s)}$, $k =$ Boltzmann constant and $T=$ temperature. This was validated by his experiments in substituting K, Rb and Cs into crystalline KCl and KNO_3 grown from the molten salt, showing that $\ln D$ varies linearly with $e^2(1+e/y)$. This theory was applied in collaboration with his colleagues N. Onuma, H. Higuchi and H. Wakita in the papers cited earlier. A similar approach was used by Iiyama (1974), and Iiyama and Volfinger (1976), Kanno (1977) who referred to it as the *local lattice deformation model*.

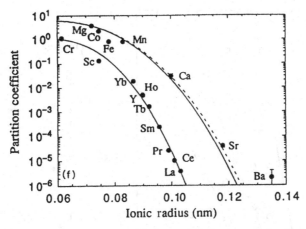

Fig. 2.15 Olivine–melt partition coefficients for divalent and trivalent cations, measured at 1190 °C, shown with curves through points calculated using the Brice theory (Beattie, 1994, Fig. 4).

The expression derived by Nagasawa (Eq. 2.54) has been superseded in more recent work (Beattie, 1994; Blundy and Wood, 1994; La Tourette *et al.*, 1995; Wood and Blundy, 1997) by a different but closely similar formula, taken from Brice (1975). Partition coefficients measured by these authors are more precise than the earlier values, so their Onuma diagrams show smoother curves (see Fig. 2.15). To calculate D_i for element i of radius r_i, at a given pressure, temperature and bulk composition, it is necessary to know D_a and r_a for an isovalent element, and then the formula given by Blundy and Wood is

$$\ln \frac{D_i}{D_a} = -\frac{4\pi E N_a}{RT}\left(\frac{r_a}{2}(r_i - r_a)^2 + \frac{1}{3}(r_i - r_a)^3\right) \qquad (2.55)$$

where E is the bulk (Young's) modulus and N_a the Avogadro number.

A similar treatment was given by Möller (1988) putting the emphasis on the close relationship between D_i and the ionic volume, or r_i^3.

2.8.2 Influence of mineral chemistry

The bulk composition of a mineral influences the ease with which a trace element can enter a site, in a different way from the effect of elastic energy just discussed.

For instance, Kretz (1959) found that the partition of V between biotite and garnet appeared quite irregular (Fig. 2.16a), but when expressed as a function of the Ca content of the garnet (Fig. 2.16b), the minerals do not appear to be out of equilibrium. This dependence on another cation substitution may be contrasted

(a)

(b)

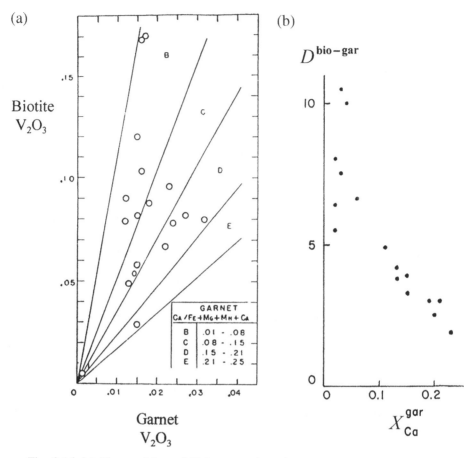

Fig. 2.16 (a) The partition of V (expressed as the oxide) between biotite and garnet, extracted from a set of high-grade gneisses, appears irregular, but is clearly influenced by the Ca content of the garnet (Kretz, 1959, Fig. 8). (b) The D-values in (a) show strong dependence on the (calculated) Ca-end-member in the garnet.

with Fig. 2.1, where V partition between biotite and amphibole is linear, with no sign of control by another element.

In a different example, Fig. 2.17 shows how the values of D_{Sr} and D_{Ba} between plagioclase and melt depend on the bulk composition of the plagioclase, as expressed by the % anorthite, and increasing with albite content.

In another example, Table 2.2 shows clinopyroxene–glass coefficients for five trace elements added as spikes to diopside–anorthite and diopside–albite mixtures, fused at appropriate temperatures at 1 bar and then cooled sufficiently to crystallise some pyroxene. The quenched pyroxene and glass were analysed by SIMS. Nineteen elements were determined and most showed rather similar distribution

Table 2.2 *Experimental partition coefficients (clinopyroxene–melt)
for trace elements in diopside–anorthite and diopside–albite
systems, showing the effects of bulk composition*[1]

	Di/An = 65/35 $D^{cpx-melt}$	Di/Ab = 55/45 $D^{cpx-melt}$	Ratio of Ds
Rb	0.000 24	0.000 21	1.17
Zr	0.2	0.045	4.57
Nb	0.006	0.002 8	2.1
Ta	0.019	0.002 4	8.11
Th	0.008	0.001 3	6.53

[1] from Lundstrom *et al.*, 1998, Table 2.

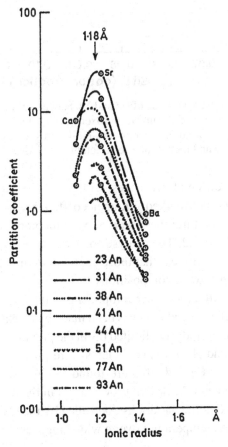

Fig. 2.17 The partition coefficients of Sr and Ba between plagioclase and silicate melt vary widely – by over an order of magnitude for Sr – according to the mineral bulk composition, expressed by the anorthite content (Jensen, 1973, Fig. 12).

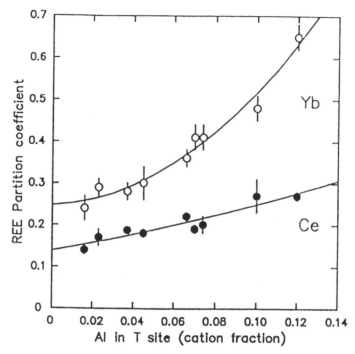

Fig. 2.18 The partition coefficients of Ce and Yb between clinopyroxene and melt vary with the proportion of Al in the pyroxene which has substituted into the T site (CaTs). These variations exceed any induced by other variables such as temperature, pressure and melt structure (Gaetani and Grove, 1995, Fig. 7a).

coefficients in the two different pyroxenes, with ratios close to unity, as in the case of Rb. Other elements had markedly higher coefficients in the Ca-rich mixture relative to the Na-rich mixture; these were mostly incompatible elements such as the four included in Table 2.2. The analyses of the glasses for these four elements are not very different so it is clear that those coefficients are largely determined by the differences in the pyroxene compositions.

A direct study of the effects of other components in the mineral on partitioning of REE between clinopyroxene and melt has been made by Gaetani and Grove (1995). This was an experimental study and limited (as in the previous example) to a system which excluded FeO and H_2O. Measurements were precise and the variations in partition coefficients for Ce and Yb were found to depend on the proportion of the Tschermak component ($CaAlSiAlO_6$ or 'CaTs') in the pyroxene (Fig. 2.18). It was found in addition that the variations controlled by the CaTs exceed variations induced by pressure and temperature over a wide range, and effects of variation in melt structure.

Similarly Lundstrom *et al.* (1994) found that the partition coefficients of Th, U and Zr between clinopyroxene and basaltic melt show a strong dependence on Al, but also on Na, in the pyroxene.

2.9 The Henry's law controversy

Since the beginning of studies of trace element substitution into minerals there were hopes that the low concentration levels would ensure that element behaviour under constant external conditions would approach ideality, and depend solely on the concentration.

As already stated, the behaviour of an ion i in any reaction or equilibrium is dependent on its activity a, which is equal to the molar concentration x multiplied by the activity coefficient γ,

$$a_i = x_i \gamma_i \tag{2.56}$$

and so, if $\gamma = 1$, then

$$a_i = x_i \tag{2.57}$$

which is the condition for ideal behaviour or conformity to Henry's law. One consequence is that from Eq. (2.6) we can write

$$\frac{a_i^\beta}{a_i^\alpha} = \frac{x_i^\beta}{x_i^\alpha} = D_i^{\beta-\alpha} = \text{constant}_{P,T} \tag{2.58}$$

or the partition coefficient is a constant for a given pressure and temperature. Now the bulk of the studies reported in the last few pages document the *failure* of Henry's law in various circumstances, i.e. D is not constant for a given P and T. But the aim of those studies has been to find an *adjustment* which would yield a coefficient invariable except to external constraints.

For a number of years, analytical methods were not available to measure D-values directly at trace levels. Concentrations of up to 5% were used to measure REE coefficients such as $D_{Sm}^{plag-melt}$ (Drake and Weill, 1975). It became of some importance to determine whether such D-values could safely be used at trace levels in modelling, so several studies in the 1970s and 1980s focused on the variation, or lack of variation, in $D_i^{min-melt}$ with the concentration c_i^{melt}, as affected by analytical errors. For instance, Drake and Weill (1975) measured REE coefficients which showed essentially constant values over a range of c_i^{melt} values from ppm to several %, using EMP analysis. B. Mysen, however, in a series of papers (e.g. 1976, 1978a, 1978b) showed non-linear variation of $D_i^{min-melt}$ with c_i^{melt}, using the rather new analytical method of beta-ray radiography. Mysen's experiments showed a decrease in $D_{Sm}^{oliv-melt}$, $D_{Sm}^{opx-melt}$, $D_{Ni}^{oliv-melt}$, $D_{Ni}^{opx-melt}$ as the concentration c_i^{melt} increased, but Harrison (1979) and Harrison and Wood (1980) found that the coefficients for Sm and Tm for garnet–melt and diopside–melt showed the reverse behaviour. Resolution of these conflicting results was attempted by repeating previous experiments (Drake and Holloway, 1981, 1982), which suggested that beta-track radiography did not give reliable results. Repeat analyses by SIMS were made of some of the

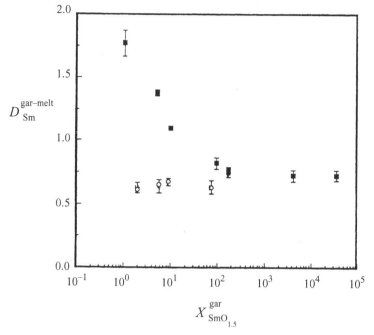

Fig. 2.19 Repeat measurements by SIMS (open circles; Beattie, 1993) of $D_{Sm}^{gar-melt}$ using samples prepared and analysed initially by beta-track mapping (filled squares; Harrison, 1979) show marked differences at the four lowest concentrations. The SIMS values are similar to the beta-track values at higher concentrations, so it is inferred that beta-track analyses are faulty at low concentration levels (Beattie, 1993, Fig. 2).

samples analysed by Harrison (1979), and the results (Beattie, 1993) show clear disagreements (see Fig. 2.19). Careful interpretation led Beattie to conclude that the beta-track method was indeed unreliable at low concentrations and thus that $D_{Sm}^{plag-melt}$ does not vary with concentration.

It seems now to have been established that D-values can in fact be measured reliably over a wide range of minor to trace concentrations, and that such values are constant for systems which do not show serious changes in other components, i.e. that in these circumstances Henry's law is valid.

2.10 Use of partition coefficients

The considerable volume of partition coefficient studies in recent years, as documented in previous paragraphs, was undertaken because of a need for good values in modelling natural processes. Attempts have thus been made to assemble lists of D-values for this purpose from time to time, such as those by Irving (1978) in the wake of the Sedona conference mentioned above, and another well-known compilation by Jones (1995). However, the non-compliance with Henry's law in multi-component systems limits the value of such lists.

Even in circumstances where D-values are known with some confidence, the use of such coefficients must be appropriate to the aim of the project. A. Treiman (1996) has illustrated this principle well by drawing attention to the difficulty of using basalt mineralogy to recover the composition of the parent magma. If the present mineralogy is an assemblage which crystallised in equilibrium from the magma, then the value $D_{REE}^{min-melt}$ permits calculating the melt REE concentration from a mineral analysis. But the present basalt mineralogy is the result of sub-solidus crystallisation, so using even the best coefficients will give false results for parent magma compositions, as Treiman shows with well-chosen examples.

2.11 Summary

The presence of an extraneous element in two minerals may often be attributed to a non-essential component or compound of the element, because minerals are not pure phases but solutions. The behaviour of such a non-essential component which is present in two minerals, or in other kinds of solutions, may be usefully discussed using the concept of the *partition coefficient*, or concentration ratio of the component between the two phases. The extraneous element in many cases occurs in site substitution for a major element, in which case its cation fraction may serve as the component concentration. The coefficient may be expressed in terms of weight concentration or molar concentration, the two being simply related by a constant term. The former is easier to calculate from analytical data.

Element partition in a system consisting of two or more phases (α, β) may be examined by treating the phases either as coexisting solutions or as reacting solutions, depending on the circumstances. In the latter case the mass-action exchange coefficient K_{i-j} with another element j is simply related to the partition coefficient $D_i^{\beta-\alpha}$ and, if i is a trace component, the two are nearly equal.

If the element of interest is present at trace concentrations then the partition coefficient may be treated as being constant over some concentration range, given a constant T and P and constant abundances of other components. Changes of temperature and pressure will produce changes in $D_i^{\beta-\alpha}$ which, if calibrated, might serve as a geothermometer or geobarometer. Variation in concentration of other components in the system may also induce changes in $D_i^{\beta-\alpha}$ which, if calibrated, might also serve as a geothermometer or geobarometer.

Developments in analytical methods, particularly EMP and SIMS, have permitted more accurate measurements of $D_i^{\beta-\alpha}$ and studies of the effects of concentration variation in other components in the system. Important effects result from major element variation, (i) of other metals substituting at the same structural site as i, (ii) of degree of silicate polymerisation and anion concentration, (iii) of volatile concentrations and fO_2.

The magnitude of a metal partition coefficient is related to, and can be calculated from, the energy required to force its cation into a structural site in a crystal. This energy depends on (i) the size and valency of the cation, (ii) other elements substituting at the same site, and (iii) the bulk composition in the crystal's solid solution series.

In systems of nearly constant composition the value of $D_i^{\beta-\alpha}$ for element i is constant, regardless of the concentration level (in the minor or trace range), subject only to variation in response to external variables such as P and T; this is conformity to Henry's law.

References

Allègre, C. J., M. Treuil, J. F. Minster, B. Minster and F. Albarede (1977) Systematic use of trace elements in igneous process. Part I: Fractional crystallization processes in volcanic suites. *Contributions to Mineralogy and Petrology*, **60**, 57–75.

Austrheim, H. and W. L. Griffin (eds.) (2000) Element partitioning in geochemistry and petrology (Special volume in honour of Brenda B. Jensen). *Lithos*, **53**, 165–292.

Barth, T. F. W. (1956) Studies in gneiss and granite. *Skrifter norske videnskaaps akademie i Oslo i math-nat. Klasse*, **1**, 387.

Bea, F. (1996) Controls on trace element composition of crustal melts. *Geological Society of America special paper*, **315**, 33–41.

Beattie, P. (1993) On the occurrence of apparent non-Henry's Law behaviour in experimental partitioning studies. *Geochimica et cosmochimica acta*, **57**, 47–56.
  (1994) What determines the values of mineral-melt partition coefficients (abstr). *Goldschmidt Conference Proceedings, Mineralogical Magazine*, **58A**, 63–4.

Beattie, P., M. J. Drake, J. H. Jones, W. Leeman, J. Longhi, G. McKay, R. Nielsen, H. Palme, D. M. Shaw, E. Takahashi and B. Watson (1993) Terminology for trace element partitioning. *Geochimica et cosmochimica acta*, **57**, 1605–6.

Beswick, A. E. (1973) An experimental study of alkali metal distributions in feldspars and micas. *Geochimica et cosmochimica acta*, **37**, 183–208.

Bindeman, I. N., A. M. Davis and M. J. Drake (1998) Ion microprobe study of plagioclase-basalt partition experiments at natural concentration levels of trace elements. *Geochimica et cosmochimica acta*, **62**, 1175–93.

Blundy, J. and B. Wood (1994) Prediction of crystal-melt partition coefficients from elastic moduli. *Nature*, **372**, 452–4.

Brice, J. C. (1975) Some thermodynamic aspects of the growth of strained crystals. *Journal of Crystal Growth*, **28**, 249–53.

Cameron, K. L. and M. Cameron (1986) Whole-rock/groundmass differentiation trends of REEs in high-silica rhyolites. *Geochimica et cosmochimica acta*, **50**, 759–70.

Cawthorn, R. G. and J. R. McIver (1977) Nickel in komatiites. *Nature*, **266**, 716–18.

Colson, R. O., G. A. McKay and L. A. Taylor (1988) Temperature and composition dependencies of trace element partitioning: olivine/melt and low-Ca pyroxene/melt. *Geochimica et cosmochimica acta*, **52**, 539–54.

Drake, M. J. and J. R. Holloway (eds.) (1978) Experimental trace element geochemistry. Proceedings of an international conference held in Sedona, Arizona, 12–16 September 1977. *Geochimica et cosmochimica acta*, **42 (6A)**, 657–943.

Drake, M. J. and J. R. Holloway (1981) Partitioning of Ni between olivine and silicate melt: the 'Henry's Law' problem reexamined. *Geochimica et cosmochimica acta*, **45**, 431–8.

(1982) Partitioning of Ni between olivine and silicate melt: the 'Henry's Law' problem reexamined: reply to discussion by B. Mysen. *Geochimica et cosmochimica acta*, **46**, 299.

Drake, M. J. and D. F. Weill (1975) Partition of Sr, Ba, Ca, Y, Eu^{2+}, Eu^{3+} and other REE between plagioclase feldspar and magmatic liquid: an experimental study. *Geochimica et cosmochimica acta*, **39**, 689–712.

Ellison, A. J. G. and P. C. Hess (1990) Two-lattice models of trace element behaviour: a response. *Geochimica et cosmochimica acta*, **54**, 2297–302.

Eugster, R. H. (1955) The Cs-K equilibrium in the system sanidine-water. *Carnegie Institute of Washington Year Book*, **54** (1954–5), 112–13.

Flynn, R. T. and C. W. Burnham (1978) An experimental determination of REE partition coefficients between a chloride-containing vapor phase and silicate melts. *Geochimica et cosmochimica acta*, **42**, 685–702.

Foley, S. F. and S. R. Van der Laan (eds.) (1994) Trace-element partitioning with application to magmatic processes. *Chemical Geology*, **117 (Nos. 1–4)**, 1–391.

Frederickson, A. F. (1962) Partition coefficients – new tool for studying geological problems. *American Association of Petroleum Geologists, Bulletin*, **46**, 518–28.

Gaetani, G. A. and T. L. Grove (1995) Partitioning of REE between clinopyroxene and silicate melt: crystal chemical controls. *Geochimica et cosmochimica acta*, **59**, 1951–62.

Grutzeck, M., H. Skirdelbaugh and D. Weill (1974) The distribution of Sr and REE between diopside and silicate liquid. *Geophysical Research Letters*, **1**, 273–6

Häkli, T. A. and T. L. Wright (1967) The fractionation of nickel between olivine and augite as a geothermometer. *Geochimica et cosmochimica acta*, **31**, 877–84.

Harrison, W. J. (1979) An experimental investigation of the partitioning of a REE, Sm^{3+}, between garnets and melt at high pressure and temperature with particular reference to non-Henry's law behavior. Ph.D. thesis, University of Manchester.

Harrison, W. J. and B. J. Wood (1980) An experimental investigation of the partitioning of REE between garnet and liquid with reference to the role of defect equilibria. *Contributions to Mineralogy and Petrology*, **72**, 145–55.

Hart, S. R., K. E. Davis, I. Kushiro and E. B. Watson (1976) Partitioning of Ni between olivine and silicate liquid. *Geological Society of America, Abstracts with Programs*, **8**, 906.

Higuchi, H. and H. Nagasawa (1969) Partition of trace elements between rock-forming minerals and the host volcanic rocks. *Earth and Planetary Science Letters*, **7**, 281–7.

Iiyama, J. T. (1970) Influence de la pression sur la composition de la solution hydrothermale sodi-potassique de divers sels en équilibre avec les feldspaths perthitiques à 600 °C. *Comptes rendus de l'académie des sciences de Paris*, Sér. D, **271**, 1925–7.

(1974) Substitution, déformation locale de la maille et équilibre de distribution des éléments en traces entre silicates et solution hydrothermales. *Bullétin de la société Française de minéralogie et de cristallographie*, **97**, 143–51.

Iiyama, J. T. and M. Volfinger (1976) A model for trace element distribution in silicate structures. *Mineralogical Magazine*, **40**, 555–64.

Irvine, T. R. and I. Kushiro (1976) *Carnegie Institute of Washington Yearbook* 75, 668–75.

Irving, A. J. (1978) A review of experimental studies of crystal/liquid trace element partitioning. *Geochimica et cosmochimica acta*, **42–6A**, 743–70.

Jensen, B. B. (1973) Patterns of trace element partitioning. *Geochimica et cosmochimica acta*, **37**, 2227–42.

Jones, J. H. (1984) Temperature- and pressure-independent correlations of olivine/liquid partition coefficients and their application to trace element partitioning. *Contributions to Mineralogy and Petrology*, **88**, 126–32.

Jones, J. H. *et al.* (1995) Experimental investigations of the partitioning of Nb, Mo, Ba, Ce, Pb, Ra, Th, Pa, and U between immiscible carbonate and silicate liquids. *Geochimica et cosmochimica acta*, **59**, 130–200.

Kanno, H. (1977) A theoretical attempt to interpret patterns of trace element partitioning in silicate minerals. *Geochemical Journal*, **11**, 155–60.

Keppler, H. (1993) Influence of F on the enrichment of HFS trace elements in granitic rocks. *Contributions to Mineralogy and Petrology*, **114**, 479–88.

  (1996) Constraints from partitioning experiments on the composition of subduction-zone fluids. *Nature*, **380**, 237–9.

Kohn, S. C. and P. F. Schofield (1994) The importance of melt composition in controlling trace element behaviour: an experimental study of Mn and Zn partitioning between forsterite and silicate melts. *Chemical Geology*, **117**, 73–87.

Kretz, R. (1959) Chemical study of garnet, biotite and hornblende from gneisses of SW Quebec, with emphasis on distribution of elements in coexisting minerals. *Journal of Geology*, **67**, 371–402.

La Tourette, T., R. L. Hervig and J. R. Holloway (1995) Trace element partitioning between amphibole, phlogopite and basanite melt. *Earth and Planetary Science Letters*, **135**, 13–30.

Leeman, W. P. (1973) Partitioning of Ni and Co between olivine and basaltic liquid: an experimental study. *EOS, Transactions American Geophysical Union*, **54**, 1222.

Lundstrom, C. C., H. F. Shaw, F. J. Ryerson, D. L. Phinney, J. B. Gill and Q. Williams (1994) Compositional controls on the partitioning of U, Th, Ba, Pb, Sr and Zr between clinopyroxene and haplobasaltic melts: implication for uranium series disequilibria in basalts. *Earth and Planetary Science Letters*, **128**, 407–23.

Lundstrom, C. C., H. F. Shaw, F. J. Ryerson, Q. Williams and J. B. Gill (1998) Crystal chemical control of clinopyroxene-melt partitioning in the Di-An-Ab system: implications for elemental fractionations in the depleted mantle. *Geochimica et cosmochimica acta*, **62**, 2849–63.

Masuda, A. (1965) Geochemical constants for Rb and Sr in basic rocks. *Nature*, **205**, 555–8.

McIntire, W. L. (1963) Trace element partition coefficients – a review of theory and applications to geology. *Geochimica et cosmochimica acta*, **27**, 1209–64.

Möller, P. (1988) The dependence of partition coefficients on differences of ionic volumes in crystal-melt systems. *Contributions to Mineralogy and Petrology*, **99**, 62–9.

Mysen, B. (1976) Partitioning of Sm and Ni between olivine, orthopyroxene and liquid: preliminary data at 20 kbar and 1025 °C. *Earth and Planetary Science Letters*, **31**, 1–7.

  (1978a) Experimental determination of Ni partition coefficients between liquid pargasite and garnet peridotite minerals and concentration limits of behavior according to Henry's Law at high pressure and temperature. *American Journal of Science*, **278**, 217–43.

(1978b) Limits of solution of trace elements in minerals according to Henry's Law: review of experimental data. *Geochimica et cosmochimica acta*, **42–6A**, 871–86.

Mysen, B. and I. Kushiro (1979) Pressure dependence of Ni partitioning between forsterite and aluminous silicate melts. *Earth and Planetary Science Letters*, **42**, 383–8.

Mysen, B. O. and D. Virgo (1980) Trace element partitioning and melt structure: an experimental study at 1 atm pressure. *Geochimica et cosmochimica acta*, **44**, 1917–30.

Nabelek, P. I. (1980) Nickel partitioning between olivine and liquid in natural basalts: Henry's Law behaviour. *Earth and Planetary Science Letters*, **48**, 293–302.

Nagasawa, H. (1966) Trace element partition in ionic crystals. *Science*, **152**, 767–9.

Nielsen, R. L. (1985) A method for the elimination of the compositional dependence of trace element distribution coefficients. *Geochimica et cosmochimica acta*, **49**, 1775–80.

(1988) A model for the simulation of combined major and trace element liquid lines of descent. *Geochimica et cosmochimica acta*, **52**, 27–38.

Onuma, N., H. Higuchi, H. Wakita and H. Nagasawa (1968) Trace element partition between two pyroxenes and the host lava. *Earth and Planetary Science Letters*, **5**, 47–51.

Philpotts, J. A. (1978) The law of constant rejection. *Geochimica et cosmochimica acta*, **42–6A**, 909–20.

Reed, M. J. (1995) Experimental determination of REE distribution between Cl-bearing volatile phase and granitic melt (abstr) in *The Origin of Granites and Related Rocks*, ed. M. Brown and P. M. Piccoli, United States Geological Survey Circular 1129, 122.

Schmitt, W., H. Palme and H. Wänke (1989) Experimental determination of metal/silicate partition coefficients for P, Co, Ni, Cu, Ga, Ge, Mo and W and some implications for the early evolution of the Earth. *Geochimica et cosmochimica acta*, **53**, 173–86.

Signorelli, S. and M. R. Carroll (2000) Solubility and fluid-melt partitioning of Cl in hydrous phonolitic melts. *Geochimica et cosmochimica acta*, **64**, 2851–62.

Sun, C. O., R. J. Williams and S. S. Sun (1974) Distribution coefficients of Eu and Sr for plagioclase-liquid and clinopyroxene-liquid equilibria in oceanic ridge basalt: an experimental study. *Geochimica et cosmochimica acta*, **38**, 1415–34.

Takahashi, E. (1978) Partitioning of Ni^{2+}, Co^{2+}, Fe^{2+}, Mn^{2+} and Mg^{2+} between olivine and silicate melts: compositional dependence of partition coefficient. *Geochimica et cosmochimica acta*, **42**, 1829–44.

Takahashi, E. and T. N. Irvine (1981) Stoichiometric control of crystal/liquid single-component partition coefficients. *Geochimica et cosmochimica acta*, **45**, 1181–6.

Treiman, A. H. (1996) Cumulate eucrites formed from normal eucritic magmas. *Lunar and Planetary Science*, **24**, 1337–8.

Treuil, M., J.-L. Joron, H. Jaffrezic, B. Villemant and B. Calas (1979) Géochimie des éléments hygromagmatophiles, coéfficients de partage minéraux/liquides et propriétés structurales de ces éléments dans les liquides magmatiques. *Bullétin de Minéralogie*, **102**, 402–9.

Webster, J. D., J. R. Holloway and R. L. Hervig (1989) Partitioning of lithophile trace elements between H_2O and H_2O-CO_2 fluids and topaz rhyolite melt. *Economic Geology*, **84**, 116–34.

Wones, D. R. and H. P. Eugster (1965) Stability of biotite: experiment, theory and application. *American Mineralogist*, **50**, 1228–70.

Wood, B. J. and J. D. Blundy (1997) A predictive model for REE partitioning between clinopyroxene and anhydrous silicate melt. *Contributions to Mineralogy and Petrology*, **129**, 166–81.

(2002) The effect of H_2O on crystal-melt partitioning of trace elements. *Geochimica et cosmochimica acta*, **66**, 3647–56.

3

Crystallisation: basic trace element modelling

3.1 Introduction

This chapter is concerned with discussing the possible ways in which a single trace element might behave during the crystallisation of a silicate melt or magma.

The initial system will consist of a single phase – a melt – at a temperature above the liquidus, intruded into or extruded onto cooler surroundings in the Earth's crust. As the melt cools various events take place: solid mineral phases begin to crystallise out; assimilation of the surrounding wall-rocks begins; expulsion of dissolved volatile components occurs; separation of a second liquid phase is possible, etc.

This chapter will focus on mineral crystallisation and the system will be treated as closed i.e. no material enters or leaves; it will also be supposed that the proportions of minerals crystallising are constant, and partition coefficients are constant too. The evolution of the system may develop under either *equilibrium* or *fractional* crystallisation. The theory for the magmatic system applies equally well to some aspects of vein formation, pegmatite growth and metamorphic grain growth, as well as to fields outside earth science such as chromatography, metal purification and distillation of brandy. The work of Lord Rayleigh (1902) on the distillation of binary mixtures provided the theory for fractional crystallisation.

Adaptations of this theory to simple geochemical systems were formulated by Doerner and Hoskins (1925) for growing crystals, Neumann (1948) for vein deposits, Holland and Kulp (1949) for pegmatites, Neumann *et al.* (1954) for igneous crystallisation, Sabatier (1971) for hydrothermal transfer, and others. A notable contribution to igneous systems was made by Greenland (1970).

The actual crystallisation of cooling magma is much more complicated than a simple division into equilibrium and fractional modes. The behaviour of trace elements will be subject to kinetic factors such as *crystal nucleation* and *growth rate*. These will not be considered here but are discussed by Marsh (1998).

3.2 Definitions

The following symbols will be used in this and succeeding chapters (it should be noted that superscripts are capitalised to indicate equilibrium conditions):

L_0	initial mass of melt or liquid
L	mass of residual liquid
$F = L/L_0$	fraction of residual liquid
W	mass of solids crystallised
W^i	mass of mineral i crystallised
w_0	mass of trace element in initial liquid
w^l or w^L	mass of trace element in residual liquid
w^s or w^S	mass of trace element in crystallised solids
w^i	mass of trace element in mineral i
c_0	initial concentration of an element
c^l or c^L	concentration of an element in residual liquid
c^s or c^S	concentration of an element in solids crystallised
c^i	concentration of an element in mineral i
$\bar{c}$	concentration of an element in accumulated solid fractions
X_i	mass fraction of mineral i among all the solid phases
D^{i-l}	partition coefficient for an element, between mineral i and liquid l
D or D^{wr}	weighted or whole rock partition coefficient
D_{i-j}	mass action exchange constant for elements i and j

3.3 Temporal variables in a crystallising system

In examining the behaviour of a trace element in a magmatic system it is necessary to consider the size and perhaps the geometry of the system and the length of time that it takes to crystallise or evolve to a certain degree. The size may be specified by mass, geometry by length coordinates and time by years. These are extensive variables, but for model studies it is frequently more useful to use dimensionless, intensive variables; one of the most useful is F, the fraction of liquid remaining. The relationship of F to time variables will now be considered.

It is assumed that a magma system evolves during some time period t at a certain rate dL/dt, where L is the mass of the residual liquid, accompanied by crystallised minerals. This rate and mass are usually unknown and so the degree of evolution, F, is used for modelling.

Then

$$F = \frac{L}{W_0} \tag{3.1}$$

and

$$\frac{dL}{L} = \frac{dF}{F} = d \ln F \qquad (3.2)$$

and so

$$d \ln F = \frac{dL}{dt} \cdot \frac{dt}{L} \qquad (3.3)$$

If now the average rate of mass change[1] $-R_c$ is substituted for dL/dt then

$$d \ln F = \frac{-R_c}{L} dt \qquad (3.4)$$

This equation may be adapted to other processes such as assimilation; if the rate constant of assimilation is $+R_a$ then

$$d \ln F = \frac{-R_c + R_a}{L} dt \qquad (3.5)$$

where $1 > F > 0$ and $0 < t < t_{end}$. The use of this relation has been explored by De Paolo (1981) in a study of the relative importance of assimilation and crystallisation.

Of course, F is poorly known in almost all natural igneous rock series. Some workers have tried to estimate values appropriate to their samples, using the known facts about the rocks. This is reverse modelling or *inversion* and will be discussed later (a simple example has been briefly described by Barca *et al.*, 1988, using mineral proportions, partition coefficients and concentrations of trace elements). In spite of the problems, the proportion of melt fraction F will be used extensively in the following sections.

3.4 Equilibrium crystallisation

In considering equilibrium crystallisation, the starting point is a hot liquid magma, which cools until some mineral becomes super-saturated and begins to crystallise. Heat continues to be lost, the temperature continues to decrease and the liquid changes composition along the *liquidus*, a *boundary curve* in the phase diagram; if the mineral is a solid solution its composition changes along a *solidus* (see Fig. 3.1). If the two phases remain in equilibrium, this implies that the rates of diffusion of elements within each phase are faster than the rate of cooling of the system.

If at some point a second mineral begins to form, heat continues to be lost and the temperature may continue to decrease or may remain constant. In the first case the liquid evolves along a *boundary curve* or *surface* and the minerals crystallise in changing proportions; in the latter case the system has reached an *invariant point*

---

[1] Mass decrease is taken as negative.

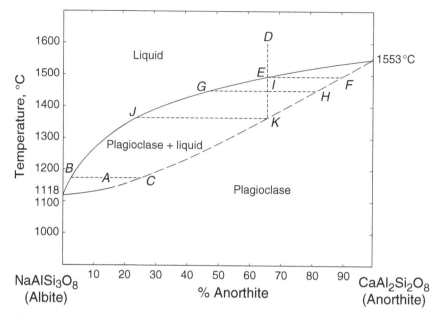

Fig. 3.1 The plagioclase binary system (after Bowen, 1913 and Ehlers, 1972) serves to illustrate the principles of simple magmatic crystallisation; a cooling homogeneous liquid (*D*) begins to crystallise when its temperature intersects the liquidus (*BJGE*) at *E*. The solid composition at this temperature is given by the solidus (*CKHF*) at *F*. If the phases are *in equilibrium*, as cooling continues, the liquid and solid compositions change along the two boundary curves. At the isotherm *JK* the last drop of liquid is used up because the composition of the solid (*K*) has become the same as the initial liquid (*D*).

and the minerals crystallise in fixed or *eutectic* proportions. Complications arise if the liquid composition reaches a field where one of the minerals becomes unstable and a *peritectic reaction* is initiated, which will be considered later.

Assuming a closed system, with one or more minerals forming in constant proportions, the trace element concentration c^L, when a proportion F of melt remains, will be governed by the *bulk* or *whole rock partition coefficient* D^{wr} as follows. If the coefficient for mineral i is D^{i-1}, and the concentrations in i and liquid are c^i and c^L then

$$c^i = c^L D^{i-1} \tag{3.6}$$

The concentration in the bulk solids being precipitated is c^S, and if the proportion of i is p_i then

$$c^S = X_1 c^1 + X_2 c^2 + \cdots = \sum_n X_i c^i = c^L \sum_n X_i D^{i-1} \tag{3.7}$$

and

$$D^{\mathrm{wr}} = \frac{c^{\mathrm{S}}}{c^{\mathrm{L}}} = \sum_n X_i D^{i-1} \qquad (3.8)$$

The mass balance of the trace element may then be written

$$w^{\mathrm{L}} = w_0 - w^{\mathrm{S}} \qquad (3.9)$$

so

$$c^{\mathrm{L}} L = c_0 L_0 - c^{\mathrm{S}} W^{\mathrm{S}} = c_0 L_0 - c^{\mathrm{L}} D^{\mathrm{wr}} W^{\mathrm{S}} \qquad (3.10)$$

and consequently, since

$$L = F W_0 \qquad W^{\mathrm{S}} = (1 - F) W_0 \qquad (3.11)$$

$$c^{\mathrm{L}} = \frac{c_0}{F + D^{\mathrm{wr}}(1 - F)} \qquad (3.12)$$

and

$$c^{\mathrm{S}} = \frac{c_0 D^{\mathrm{wr}}}{F + D^{\mathrm{wr}}(1 - F)} \qquad (3.13)$$

It is evident that the behaviour of the element is governed by the magnitude of the bulk partition coefficient, as is seen from Fig. 3.2. If $D^{\mathrm{wr}} > 1$ the liquid concentration and the solid concentration both decrease as the proportion of solids increases, whereas if $D^{\mathrm{wr}} < 1$ the opposite takes place; for instance when $D^{\mathrm{wr}} = 0.1$, the initial value of the solid concentration c^{S}/c_0 also is 0.1, and the liquid concentration ratio c^{L}/c_0 becomes asymptotic to the value 10 as F approaches zero. In both cases the relative concentration in the solid reaches unity at the end of crystallisation, because the system is closed.

A process of this kind is often referred to as *batch crystallisation*.

3.5 Fractional crystallisation

The preceding section was based on the premise that crystallised minerals remain in equilibrium with the residual magma, which implies that the diffusion of ions in the solids, to achieve re-equilibration, takes place faster than the rate of crystallisation. It may be, however, that the minerals are unable to react with residual liquid, in which case *fractional crystallisation* takes place. In Fig. 3.1, for example, if the successive solid phases *F, H, K*, etc. do not equilibrate with their liquids (perhaps by forming crystal zones enclosing earlier fractions) then the melt continues to change composition along *EGJB* and finally solidifies to pure albite at 1118 °C.

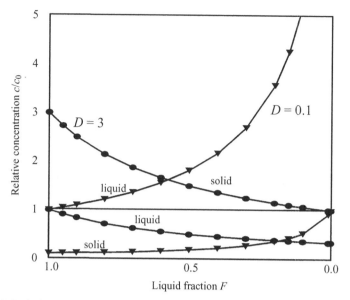

Fig. 3.2 Relative concentration (c/c_0) of a trace element in a solid phase which is crystallising in equilibrium from a liquid, as the liquid fraction decreases from 1.0 towards zero, following Eq. (3.3). If $D > 1.0$, c^s is high initially and then decreases, ending at unity, whereas $c^L/c_0 = 1$ initially and then also decreases; when $D < 1.0$ the opposite occurs.

Fractional crystallisation may be described following an approach first used by Rayleigh (1902).

A melt has mass L_0 and trace element concentration c_0. If the mass of trace element is w, then at any point during the crystallisation

$$w^l = c^l L \tag{3.14}$$

After a short period of crystallisation

mass of residual liquid is $L - dL$
mass of element in liquid is $w^l - dw^l$

or

mass of crystallised liquid is dL
mass of element lost by liquid is dw^l

So the concentration in the solid crystallising instantaneously during this short period is c^s, where

$$c^s = \frac{dw^l}{dL} = \frac{d(c^l L)}{dL} \tag{3.15}$$

Now, assume c^s is some function of c^l, say $c^s = f(c^l)$.

So

$$f[c^l] = \frac{d(c^l L)}{dL} \tag{3.16}$$

whence

$$\frac{dc^l}{dL} = \frac{f[c^l] - c^l}{L} \tag{3.17}$$

So

$$\frac{dc^l}{f[c^l] - c^l} = \frac{dL}{L} = \frac{dF}{F} \tag{3.18}$$

because $L = L_0 F$.

This may be called the *Rayleigh equation*, but if there is evidence that the relationship between c^s and c^l is linear over some range,[2] then we can write

$$c^s = Dc^l \tag{3.19}$$

and Eq. (3.18) can be integrated to give the expression

$$\frac{c^l}{c_0} = F^{D-1} \tag{3.20}$$

An alternative derivation comes from the two mass balance equations, for the system as a whole and for the trace element, which may be written

$$L_0 = L + W^s = FL_0 + (1 - F)L_0 \tag{3.21}$$

$$dL + dW^s = 0 \quad dW^s = -dF \tag{3.22}$$

$$w_0 = w^l + w^s \quad dw^l + dw^s = 0 \tag{3.23}$$

Now define c^s as the instantaneous concentration of the trace element in the crystals precipitated.

Then

$$c^s = \frac{dw^s}{dW^s} \quad D = \frac{c^s}{c^l} \tag{3.24}$$

so

$$dw^s = c^s dW^s = -c^l DL_0 dF = -\frac{w^l DL_0 dF}{L_0 F} \tag{3.25}$$

---

[2] This is evidently not the case in Fig. 3.1.

and consequently

$$dw^1 + ds^s = 0 = dw^1 - \frac{w^1 D L_0 dF}{L_0 F} \tag{3.26}$$

and

$$\frac{dw^1}{w^1} = D \cdot \frac{dF}{F} \tag{3.27}$$

so, on integration

$$\ln \frac{w^1}{w_0} = D \ln F \tag{3.28}$$

Substituting

$$w_0 = c_0 L_0 \quad w^1 = c^1 L = c^1 L_0 F \tag{3.29}$$

gives

$$\ln \left(\frac{c^1}{c_0} \cdot F \right) = D \ln F \tag{3.30}$$

or

$$\frac{c^1}{c_0} = F^{D-1}$$

which is Eq. (3.20).

The behaviour of liquid and solid concentrations in fractional crystallisation is shown in Fig. 3.3. At first sight, this diagram looks similar to Fig. 3.2, since it uses the same partition coefficients, but as the liquid fraction decreases the fractionation concentrations diverge markedly from the equilibrium ones.

If the partition coefficient is markedly greater than unity the trace element is strongly concentrated in early minerals, the melt being impoverished as a consequence; this is the behaviour of a *compatible element*. In this way the accumulation of Ni in early basaltic minerals, which are rich in the compatible major elements Mg and Ca is accounted for.

In the opposite case, when D is very much less than unity, the element shows no tendency to enter early minerals and is said to be *incompatible*. It will therefore accumulate in later fractions of the melt, along with the major elements Na, K and Si. Some such trace elements have large ionic radii, explaining the alternative name of *large-ion lithophile* or LIL elements; others are quite small, but have high charge and are labelled *high field strength* or HFS elements.

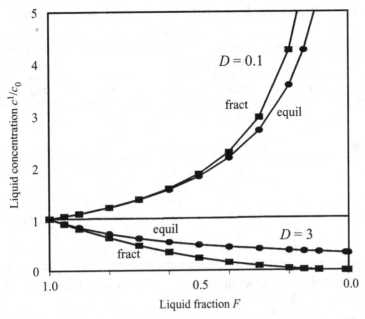

Fig. 3.3 Trace element concentrations in melts during fractional crystallisation, using the same partition coefficients as in the previous figure, compared with the equilibrium concentrations. For a compatible element ($D > 1$), the initial solid is more enriched in the trace element, and so the melt concentration (Eq. 3.20) decreases more rapidly in the fractionating liquid. For an incompatible element ($D < 1$) neither concentration changes greatly until most of the liquid has been used up, whereupon the fractionation increases strongly.

It may be noted that in the extreme case, where $D \approx 0$, then Eq. (3.20) becomes

$$\frac{c^{\mathrm{l}}}{c_0} \approx \frac{1}{F} \tag{3.31}$$

and virtually none of the element goes into the solid minerals and the liquid concentration increases towards infinite values. Such behaviour, where $D \ll 1$, has been recognised as quite common and the term *éléments hygromagmatophiles* or HM was proposed (Treuil and Joron, 1975; Treuil *et al.*, 1979).

Modelling the course of fractional crystallisation with Eq. (3.20) is interrupted if the numbers and proportions of the minerals on the liquidus change. For instance, Fig. 3.4 shows the magma concentration behaviour of Ni and Sr while a basic melt crystallises successively to peridotite, gabbro, diorite and granophyre, using the parameters given in Table 3.1.

Table 3.1 *Parameters for modelling hypothetical variation diagrams for Ni and Sr*

Minerals	Model Per Cent				Partition Coefficient	
	Perid	Gab	Dio	Gran	D_{Ni}	D_{Sr}
Spinel	5	5	5		20	0.01
Olivine	50	10			10	0.01
Hypersthene		15			1	0.01
Augite	40	30	15	3	1.5	0.5
Plagioclase	5	40	50	30	0.01	3
Quartz			10	25	0.001	0.01
Microcline			15	40	0.01	2
Zircon			5		0.001	0.001
D_{Ni}^{wr}	6.6	2.6	1.23	0.45		
D_{Sr}^{wr}	0.36	1.35	1.88	1.72		

Notes: D is the whole rock coefficient for the given mineral assemblage. Perid = peridotite, Gab = gabbro, Dio = diorite, Gran = granophyre.

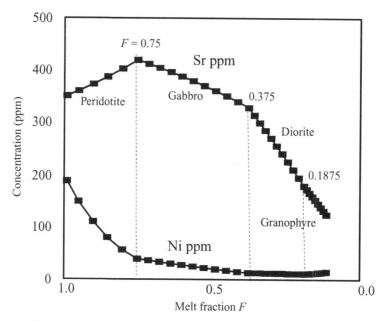

Fig. 3.4 Variation diagram for Ni and Sr fractionation during crystallisation of a mafic magma ($c_0^{Ni} = 200$ ppm, $c_0^{Sr} = 350$ ppm) to four successive mineral assemblages, in proportions given in Table 3.1. Of the initial magma, 25% crystallises to peridotite, then 50% of the residue to gabbro, then 50% of the residue to diorite, followed by granophyre.

3.6 Mineral zonation

It is assumed in fractional crystallisation that successive solid fractions are essentially isolated from the system and cannot react with the residual melt. This could occur if successive crystals sink in a less dense melt, or float in a denser melt, or grow as zones around earlier crystals. The last is sometimes referred to as a Doerner–Hoskins process, from their study (1925) of co-precipitation of Ra with barium sulphate.[3]

It is appropriate to ask what will be the bulk trace element concentration $\bar{c}$ in the accumulated solids which have formed by this process, and which are zoned crystals. Then $\bar{c}$ will be equal to the ratio of the mass of the trace element in the solid to the mass of the mineral, i.e.

$$\bar{c} = \frac{w^s}{W^s} \tag{3.32}$$

but the trace element mass in the mineral fractions is equal to the mass in the whole system *less* the mass in the liquid, that is

$$w^s = w_0 - w^l \tag{3.33}$$

so

$$\bar{c} = \frac{c_0 L_0 - L_0 F \cdot c_0 F^{D-1}}{L_0 (1 - F)} = c_0 \cdot \frac{1 - F^D}{1 - F} \tag{3.34}$$

This equation will apply to a magma crystallising to a single mineral, or an assemblage with coefficient D^{wr}. Figure 3.5 shows the accumulated relative concentration in terms of the degree of evolution of the system (Eq. 3.34) for four values of the partition coefficient. It is seen that $\bar{c}$ approaches unity as F decreases towards zero.

In the case of a mineral i, forming a constant proportion p_i in the assemblage crystallising, then its accumulated concentration is related to that for the whole assemblage by

$$\bar{c} = \sum_i p_i \bar{c}_i \tag{3.35}$$

Also, it is clear that

$$c^s = D^{wr} c^l \tag{3.36}$$

and

$$c_i^s = D^{i-1} c^l \tag{3.37}$$

---

[3] They describe the process in the following words: 'As long as crystal growth continues, radium sulfate will be buried in the larger crystals, and fresh surfaces of barium sulfate will be exposed for continued deposition'.

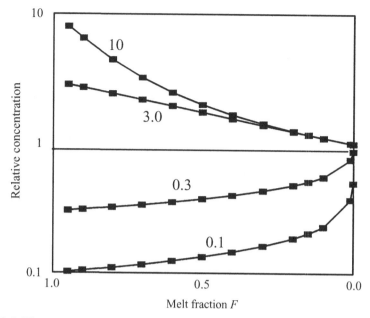

Fig. 3.5 The trace element concentration in the accumulated mineral fractions during fractional crystallisation. For compatible elements, i.e. D-values 10 and 3, the trend leads to the bulk concentration c_0 in the whole system, when all melt has solidified. For incompatible elements (D-values 0.1 and 0.3) this is also the case, but only at absurdly small melt proportions.

so

$$c_i^s = \frac{D^{i-1}}{D^{wr}} \cdot c^s \tag{3.38}$$

and this relationship applies also to the accumulated concentrations. It has already been mentioned that partition coefficients have been measured as the concentration ratio between phenocrysts and the matrix in volcanic rocks. Such measurements assume that the matrix has the composition of the magma melt, and that the phenocryst was in equilibrium with the melt throughout its crystallisation. If by contrast, however, a Doerner–Hoskins process had taken place and successive growth shells of crystal i had effectively isolated interior zones from further reaction, then the *apparent partition coefficient*, d_i, could be grossly in error. Since d_i is measured as the ratio between the average concentration throughout the crystal, which is $\bar{c}_i$, and the concentration, c^l, in the melt, Eqs. (3.20), (3.34) and (3.38) show that

$$d_i = \frac{D^{i-1}}{D^{wr}} \cdot \frac{\left(1 - F^{D^{wr}}\right)}{(1-F)F^{(D^{wr}-1)}} \tag{3.39}$$

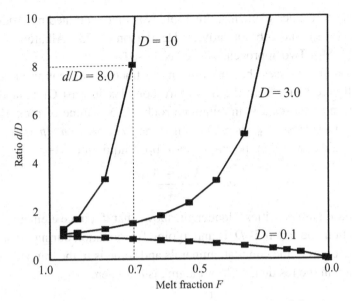

Fig. 3.6 Variation of the ratio d/D between the apparent and the true partition coefficients for a trace element, during fractionation, in relation to the fraction of residual melt. If the true coefficient is 10, the apparent coefficient is 80 after 30% crystallisation. In the early stages of crystallisation the effect is most marked for compatible elements, since the element accumulates faster in the solid phase.

Of course if only one mineral is crystallising, then D^{wr} is identical to D^{i-1} and, consequently

$$d = \frac{1 - F^D}{(1 - F)F^{D-1}} \tag{3.40}$$

The ratio d/D is shown in Fig. 3.6, which illustrates how great are the possibilities of error in using phenocrysts to estimate D^{i-1}. For instance, if the correct coefficient is 10, then the zoned crystal formed after 30% solidification would show a ratio value of 8.0, or an erroneous partition coefficient of 80.

3.7 Intercumulus trapped melt

It has been assumed up to now that, during crystallisation of the melt, the precipitated solids separate cleanly from the residual liquid. It is more likely, however, that this is not the usual case and that the minerals will coalesce to form the *cumulus phases* or *cumulates*, with some melt filling in the porosity, as *intercumulus trapped melt* or, as it is sometimes called, the *mesostasis*. If this liquid remains trapped within the solid aggregate then it will affect the subsequent behaviour of the system, by altering the mass balance of both major and trace elements.

With respect to trace elements, this topic was first examined by Anderson and Greenland (1969), and subsequently by Henderson (1975), Albarède (1976) and Langmuir (1989). Two approaches have been used.

Henderson (1975) uses the variation in the trace U concentration along a diamond drill core taken from the Skaergard complex in East Greenland to study the proportion of mesostasis in cumulate rocks. If a volume of rock V_r includes a volume of mesostasis V_m, then he defined the *fractionation efficiency* e_m as the fraction of solid material in the cumulate-trapped liquid mixture

$$e_m = \frac{V_r \rho_r - V_m \rho_m}{V_r \rho_r} \tag{3.41}$$

where ρ_r and ρ_m are densities. Henderson shows that if a trace element, such as U, whose partition coefficient is D, is undergoing fractionation during crystallisation, then some accrues to the cumulate minerals and some is in the trapped liquid. The concentration in the residual melt or magma is c^1, where

$$\frac{c^1}{c_0} = F^{e_m(D-1)} \tag{3.42}$$

An identical expression is given by Maaloe (1976). Since the degree of fractionation is F then it is readily shown that the fraction of residual melt is F_m, where

$$F_m = 1 - \frac{1-F}{e_m} \tag{3.43}$$

Henderson illustrates these relationships with the example of Ni behaviour during crystallisation of olivine where $D_{Ni}^{oliv-melt} = 16$ and the proportions of mesostasis (trapped liquid) and minerals are 0.3 and 0.7 so that

$$\frac{c^1}{c_0} = F^{0.7(16-1)} = F^{10.5} \tag{3.44}$$

The approach used by Anderson and Greenland (1969) and later by McCarthy and Robb (1978) and Albarède (1976) is to consider the trapped liquid as an additional phase, when the concentration equation becomes

$$\frac{c^1}{c_0} = F^{[xD^m+(1-x)D-1]} \tag{3.45}$$

where x and D^m are the proportion and coefficient for the mesostasis. Assuming that the mesostasis is identical to the residual liquid outside the cumulates, then D^m is a partition coefficient of unity, and the equation simplifies to

$$\frac{c^1}{c_0} = F^{[(1-x)(D-1)]} \tag{3.46}$$

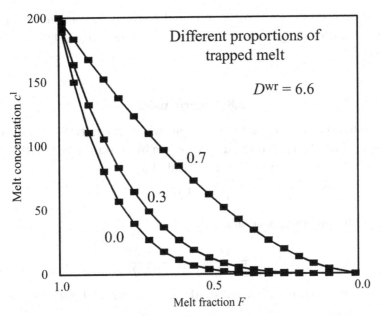

Fig. 3.7 Melt concentration of Ni during fractional crystallisation ($c_0^{Ni} = 200$) of the peridotite melt in the first stage of Fig. 3.4, with the proportions 0.0, 0.3 and 0.7 of trapped melt or mesostasis, showing that the more the trapped melt, the less the fractionation.

Continuing with the same numerical example, the coefficient for the aggregate cumulate plus trapped liquid is made up of

0.7 minerals with $\quad D_{Ni}^{oliv-melt} = 16$
0.3 mesostasis with $D_{Ni}^{melt-melt} = 1$

so that $\qquad\qquad D^{mixture} = 0.7 \times 16 + 0.3 \times 1 = 11.5$

So, in general, Eqs. (3.42) and (3.46) show that the effect of trapped melt on trace element concentrations will be to reduce the degree of fractionation, and this will apply to both compatible and incompatible elements. This is illustrated in Fig. 3.7, developed out of Fig. 3.4.

It may be added that the foregoing discussion can only apply to cases where fractional crystallisation is taking place; if equilibrium conditions apply, then the trapped liquid, cumulates and external liquid will all equilibrate with each other.

The principle that melt may be trapped in other ways during precipitation of solid phases was set forth by Langmuir (1989) in an article investigating the idea of a *solidification zone* in a cooling magma. Instead of a clear demarcation between wholly liquid melt and a cumulate zone comprising mineral aggregates and trapped melt, the solidification zone is the focus of active crystallisation, and residual liquid may move both into the cumulate zone and back into the main magma chamber. These additional degrees of freedom lead to more complicated concentration fluctuations.

3.8 Mineral pairs

The concentration ratios for a trace element which enters two or more minerals present some interesting features, as was shown by Haskin *et al.* (1970). The concentration in mineral i is, from Eqs. (3.6) and (3.38)

$$c_i^S = c^L D^{i-1} \tag{3.47}$$

during equilibrium crystallisation and

$$\bar{c_i^S} = \frac{D^{i-1}}{D} \cdot c_0 \cdot \frac{1 - F^D}{1 - F} \tag{3.48}$$

during fractional crystallisation and similarly for mineral j. It follows that

$$\frac{\bar{c_i}}{\bar{c_j}} = \frac{c_i^S}{c_j^S} = \frac{D^{i-1}}{D^{j-1}} = D^{i-j} \tag{3.49}$$

This equation says that the ratio of the *total* mineral concentrations during fractionation, is equal to the ratio of their *equilibrium* concentrations and to the ratio of the mineral–melt partition coefficients, or the partition coefficient between the two phases. In other words, the *apparent D*-value between two solid phases i and j is the true value, regardless of whether the minerals are zoned or not. This assumes, however, that the two minerals crystallised over the same period, and the system was closed.

3.9 Incremental or stepped crystallisation

At a volcanic centre lava-flows may erupt frequently, coming from an evolving magma chamber or reservoir. In some cases successive flows are nearly identical but, if the magma is crystallising this may not be the case and *incremental* changes may be observed in successive flows or *steps* in the eruptive sequence. Successive pulses of magma intrusion may similarly contribute to the generation of plutonic igneous bodies. The effect of such processes on trace element concentrations has been followed by many workers, including McCarthy and Hasty (1976), working on the granites of the Bushveld complex.

It will be assumed that the crystallised minerals become separated from melt on each step, by filter-pressing, gravitational separation or some other process. The

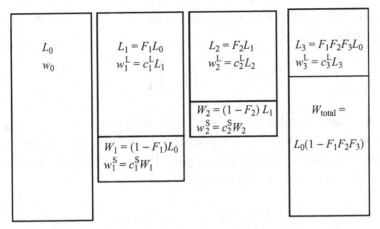

Fig. 3.8 Batch or incremental crystallisation. In step 1 the magma mass L_0 crystallises in equilibrium until a proportion F_1 is left, carrying a trace element mass w_1 and concentration c_1^L. The mineral mass W_1 formed has trace element mass and concentration w_1^S and c_1^S, and now is isolated from further interaction with the magma. The residual magma L_1 becomes the starting point for step 2, and similarly for succeeding steps. At the end of step 3 the accumulated solids are $W_1 + W_2 + W_3$, containing $w_1^S + w_2^S + w_3^S$ of the trace element, whose aggregate concentration is $\bar{c}$. Although the crystallisation on each step is an *equilibrium* or *batch* process the result of several steps is one of fractionation.

residual magma becomes the starting point for the next step, or the next increment of crystallisation. During the first step the initial magma L_0 diminishes to a fraction F_1 (see Fig. 3.8), so that $L_1 = F_1 L_0$. If samples of the successive liquids and solids can be found the process can be documented.

The mass balance of melt and crystals, and of the trace element of interest may be written

$$\text{step 1:}\quad L_0 = L_1 + W_1 \quad L_1 = F_1 L_0 \quad W_1 = (1 - F_1)L_0 \tag{3.50}$$

$$w_0 = w_1^l + w_1^s \tag{3.51}$$

$$\text{step 2:}\quad L_1 = L_2 + W_2 \quad L_2 = F_2 L_1 = F_1 F_2 L_0 \tag{3.52}$$

$$W_2 = (1 - F_2)L_1 = F_1(1 - F_2)L_0 \tag{3.53}$$

$$w_1^l = w_2^l + w_2^s \tag{3.54}$$

This formulation may be extended to subsequent steps so that, if the symbol $\prod$ be used to indicate successive multiplications, then

$$L_n = \prod_{i=1}^{n} F_i \cdot L_0 \quad W_n = (1 - F_n) \prod_{i=1}^{n-1} F_i \cdot L_0 \tag{3.55}$$

The fraction of residual liquid after the nth step is $\prod_{i=1}^{n} F_i$, and the mass of all the accumulated solids is

$$W_{\text{total}} = L_0 \left(1 - \prod_{i=1}^{n} F_i \right)$$

(3.56)

It is usually assumed that such step crystallisation is of brief duration and may therefore follow an *equilibrium* or *batch* process, in which case the element concentration at the end of the first step is given by Eq. (3.12), and this is also the starting concentration for the second step, so that

$$c_1^L = \frac{c_0}{F_1(1 - D_1) + D_1} \qquad c_2^L = \frac{c_1^L}{F_2(1 - D_2) + D_2}$$

(3.57)

and in general

$$c_n^L = \prod_{i=1}^{n} \left[\frac{c_{n-1}^L}{F_i(1 - D_i) + D_i} \right]$$

(3.58)

and

$$c_n^S = D_n c_n^L$$

(3.59)

3.9.1 Constant melt proportion

McCarthy and Hasty (1976) limited their study to cases where the residual liquid fraction, F_i, on each step was constant. Then the residue after n steps becomes F^n. The examples they use also assume that minerals crystallise in equilibrium in constant proportions, so that the partition coefficient remains constant from step to step. These two constraints simplify Eq. (3.58).

The melt concentrations after twenty steps for a partition coefficient of 0.1 are shown in Fig. 3.9a, for three values of the melt fraction on each step. The effect of 1% crystallisation on each step produces very little change in c^L over twenty steps. A 10% increment, however, produces a more marked change; the relative melt concentration on step twenty is 6.59. Similar concentrations are shown in Fig. 3.9b, plotted against the aggregate degree of crystallisation, which reaches 87.8% after twenty steps.

If the solid fraction is very small then the trace element concentration variation is essentially the same as in fractional crystallisation; for example, with $F = 0.90$, a D-value of 0.1 and a fraction 87.8 of crystallisation the relative melt concentration by fractional crystallisation will be 6.64, very similar to the previous paragraph.

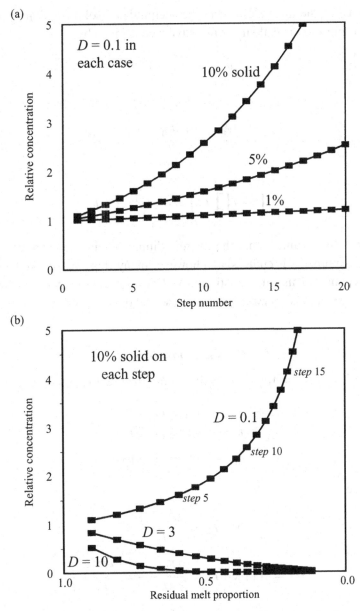

Fig. 3.9 Batch crystallisation with constant melt proportion on each step: (a) melt concentration of an incompatible trace element at the end of each of twenty steps of equilibrium crystallisation, with formation of 10, 5 and 1% solids on each step. With only 1% increment the melt concentration has hardly changed on the twentieth step; (b) The melt concentration is plotted against the residual melt, for three values of the partition coefficient with 10% crystallisation on each step. The trend for $D = 0.1$ is very similar to the trend for fractional crystallisation shown in Fig. 3.3 (see text).

If the mass of the trace element in the accumulated solid fractions is w_{total} the overall concentration is $\bar{c}$, then the mass balance is given by

$$w_{total} = w_0 - w_n^L \tag{3.60}$$

or

$$\bar{c} \cdot W_{total} = c_0 L_0 - c_n^L L_n \tag{3.61}$$

and so

$$\bar{c} \cdot \left(1 - \prod_{i=1}^{n} F_i\right) = 1 - c_n^L \cdot \prod_{i=1}^{n} F_i \tag{3.62}$$

Of course, by assuming that the magma diminishes in each step by the same fraction F_i the process is endless, and however many steps are invoked some melt will always remain. This is unrealistic and McCarthy and Hasty modified the process to one of *constant mass crystallisation* on each step.

3.9.2 Constant mass increments

In this case $W_1 = W_2 = W_3 = \cdots = W_n$. From the previous section

$$L_1 = F_1 L_0 \quad W_1 = (1 - F_1)L_0 \tag{3.63}$$
$$L_2 = F_2 L_1 \quad W_2 = (1 - F_2)L_1 \tag{3.64}$$
$$L_3 = F_3 L_2 \quad W_3 = (1 - F_3)L_2 \tag{3.65}$$

and so, since

$$W_1 = W_2 \text{ then } F_2 = \frac{2F_1 - 1}{F_1} \tag{3.66}$$

$$W_2 = W_3 \text{ then } F_3 = \frac{2F_2 - 1}{F_2} = \frac{3F_1 - 2}{2F_1 - 1} \tag{3.67}$$

$$W_3 = W_4 \text{ then } F_4 = \frac{2F_3 - 1}{F_3} = \frac{4F_1 - 3}{3F_1 - 2} \tag{3.68}$$

As a consequence, the liquid mass at the end of the nth step is given by n subtractions of a constant mass $(1 - F_1)$ g and is

$$L_n = L_0 [1 - n(1 - F_1)] \tag{3.69}$$
$$F_{resid} = 1 - n(1 - F_1) \tag{3.70}$$

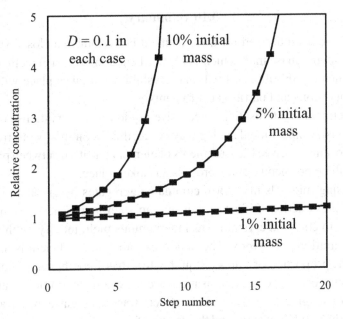

Fig. 3.10 Batch crystallisation with constant mass increment on each step. The melt concentrations show the same pattern as in the previous figure, but change more rapidly. The curves for 10%, 5% and 1% initial mass loss end respectively at step 10, step 20 and step 100.

Also, since the mass crystallised on each step is $L_0(1 - F_1)$ the maximum number of steps until no magma is left is

$$n_m = \frac{L_0}{L_0(1 - F_1)} = \frac{1}{1 - F_1} \tag{3.71}$$

The trace element concentration c_1 at the end of step 1 is given, as in the previous section, by Eq. (3.12). If we assume that the partition coefficient remains constant, as in the previous case, then by substituting for F_2 from above

$$c_2^L = \frac{c_1^L}{F_2(1 - D) + D} \tag{3.72}$$

and

$$c_n^L = \frac{c_{n-1}^L}{F_n(1 - D) + D} \tag{3.73}$$

The practical effect of this process may be seen by comparing Fig. 3.10 with Fig. 3.9b. The two processes give very similar results but the fractionation is more pronounced in a crystallisation with a constant mass increment on each step.

Repetitive or incremental crystallisation will be considered again in Chapter 5.

3.10 Summary

Equations have been derived for trace element behaviour in a closed system consisting of a single phase melt which cools until encountering a eutectic or cotectic liquidus, when equilibrium or fractional crystallisation takes place with constant mineral proportions and partition coefficients.

If, in fractional crystallisation, successive liquid residua crystallise over earlier mineral grains, compositional zoning may ensue; this is commonly referred to as the Doerner–Hoskins process. Measurements of element partition between phenocrysts and melt will be erroneous if this process has taken place.

Crystallising minerals may form cumulate aggregates by gravitational settling or by filter-press action. Silicate liquid within such aggregates may not escape, but continue to cool and crystallise as intercumulus melt, modifying the trace element concentration in the rock. The bulk rock partition coefficient is modified by including the proportion of trapped liquid, which has a coefficient of unity.

Incremental or stepped crystallisation occurs when intermittent intrusions or extrusions of magma take place. This affects trace element concentrations as a function of the number of steps and the degree of crystallisation on each, in addition to the partition coefficient.

References

Albarède, F. (1976) Some trace element relationships among liquid and solid phases in the course of the fractional crystallization of magmas. *Geochimica et cosmochimica acta*, **40**, 667–74.

Anderson, A. T. and L. J. Greenland (1969) Phosphorus fractionation diagram as a quantitative indicator of fractional crystallization differentiation of basaltic liquids. *Geochimica et cosmochimica acta*, **33**, 493–506.

Barca, D., G. M. Crisci and G. A. Ranieri (1988) Further developments of the Rayleigh equation for fractional crystallization. *Earth and Planetary Science Letters*, **89**, 170–2.

Bowen, N. L. (1913) The melting phenomena of the plagioclase feldspars. *American Journal of Science*, **35**, 577–99.

   (1928) *The evolution of the igneous rocks.* Princeton: Princeton University Press, 334p.

De Paolo, D. J. (1981) Trace element and isotopic effects of combined wallrock assimilation and fractional crystallization. *Earth and Planetary Science Letters*, **53**, 189–202.

Doerner, H. A. and W. M. Hoskins (1925) Coprecipitation of Ra and Ba sulfates. *Journal of the American Chemical Society*, **47**, 662–75.

Ehlers, E. G. (1972) *The Interpretation of Geological Phase Diagrams.* San Francisco: W. H. Freeman.

Greenland, L. P. (1970) An equation for trace element distribution during magmatic crystallisation. *American Mineralogist*, **55**, 455–65.

Haskin, L. A., R. O. Allen, P. A. Helmke *et al.* (1970) Rare earths and other trace elements in Apollo 11 lunar samples. Proceedings First Lunar Science Conference, *Geochimica et cosmochimica acta supplement*, **1**, 1213–31.

Henderson, P. (1975) Geochemical indicator of the efficiency of fractionation of the Skaergaard intrusion, East Greenland. *Mineralogical Magazine*, **40**, 285–91.

Holland, H. D. and J. L. Kulp (1949) The distribution of accessory elements in pegmatites. I. Theory. *American Mineralogist*, **34**, 35–60.

Langmuir, C. H. (1989) Geochemical consequences of *in situ* crystallization. *Nature*, **340**, 199–205.

Maaloe, S. (1976) Quantitative aspects of fractional crystallization of major elements. *Journal of Geology*, **84**, 81–96.

Marsh, B. D. (1998) On the interpretation of crystal size distributions in magmatic systems. *Journal of Petrology*, **39**, 553–99.

McCarthy, T. S. and R. A. Hasty (1976) Trace element distribution patterns and their relationship to the crystallization of granitic melts. *Geochimica et cosmochimica acta*, **40**, 1351–8.

McCarthy, T. S. and L. J. Robb (1968) On the relationship between cumulus mineralogy and trace and alkali element chemistry in an Archean granite from the Barberton region, South Africa. *Geochimica et cosmochimica acta*, **42**, 21–6.

Neumann, H. (1948) On hydrothermal differentiation. *Economic Geology*, **43**, 77.

Neumann, H., J. Mead and C. J. Vitaliano (1954) Trace element variation during fractional crystallization as calculated from the distribution law. *Geochimica et cosmochimica acta*, **6**, 90–9.

Rayleigh, Lord J. W. S. (1902) On the distillation of binary mixtures. *Philosophical Magazine*, Series 6, **IV**, 521–37.

Sabatier, G. (1971) Théorie de la distribution d'un élément en trace entre solution hydrothermale et solution solide. Application à la distribution du cesium dans la série des feldspaths alcalins. *Bullétin de la société française de minéralogie et de cristallographie*, **94**, 451–5.

Treuil, M., J.-L. Joron (1975) Utilisation des éléments hygromagmatophiles pour la simplification de la modélisation quantitative des processus magmatiques. Exemples de l'afar et de la dorsale médioatlantique. *Società Italiana di mineralogia e petrologia, Rendiconti*, **XXXI**, 125–74.

Treuil, M., J.-L. Joron, H. Jaffrezic, B. Villemant and B. Calas (1979) Géochimie des éléments hygromagmatophiles, coéfficients de partage minéraux/liquides et propriétés structurales de ces éléments dans les liquides magmatiques. *Bullétin de minéralogie*, **102**, 402–9.

4

Crystallisation: variation of mineral proportions, partition coefficients and fluid phase proportion

4.1 Introduction

The discussion so far has considered behaviour of a single trace element during silicate–melt crystallisation as a result of fractionation and of equilibrium processes, using only one variable, the proportion of residual melt. Beginning with this chapter the effects of other variables will be considered.

Among other variables, Albarède and Bottinga (1972) have reviewed the effects of kinetic disequilibrium between crystals and melt, assumptions about the size of the magma reservoir and diffusion rates of cations within the minerals and melt. They were concerned in particular with the measurement of partition coefficients from phenocrysts in lavas, and they point out that Rayleigh fractionation is only approached for rapidly diffusing elements during slow cooling. Slowly diffusing elements cause liquid heterogeneity adjacent to growing crystals, and 'trace element partition calculations . . . involving liquid and crystals require a knowledge of diffusion constants and crystal growth rates' (Albarède and Bottinga, 1972, p.154). Although this is undoubtedly true, results in succeeding years suggest that modelling calculations which neglect kinetic factors can still be useful. A number of other factors omitted from discussion so far include:

   (i) variation in proportions of minerals crystallising;
  (ii) variable partition coefficients;[1]
 (iii) volatile release during crystallisation;
 (iv) resorption of early minerals;
  (v) assimilation of wall-rock;
 (vi) periodic recharge of the magma chamber;
(vii) loss of magma by eruption or intrusion.

The first three of these will now be examined.

---

[1] Variation in partition coefficients during crystallisation was not discussed in Chapter 3.

4.2 Variation in mineral proportions

Greenland (1970) showed how variation in the proportions of minerals crystallising and variation in partition coefficients could be built into the mass balance equation for Rayleigh fractionation.

This section will consider mineral proportions, assuming partition coefficients remain constant. Using the terminology of Section 3.2, the mass balance for a crystallising magma and a trace element which it contains is given by

$$dW + dL = 0 \tag{4.1}$$

$$dw + dw^l = 0 \tag{4.2}$$

$$dw^l = d(c^l L) \tag{4.3}$$

$$dw = c^s dW \tag{4.4}$$

where dw is the instantaneous mass of the trace element in the crystallising solid dW. If the concentration in one of the minerals making up the solid is c_i, then

$$dw^i = c^i dW^i \quad dw = \sum dw^i \tag{4.5}$$

$$dw = \sum c^i dW^i = c^l \sum D^{i-l} dW^i \tag{4.6}$$

It follows that

$$c^l \sum D^{i-l} dW^i + L dc^l + c^l dL = 0 \tag{4.7}$$

or

$$\frac{dc^l}{c^l} + \frac{dF}{F} + \frac{\sum D^{i-l} dW^i}{L} = 0 \tag{4.8}$$

but since $W^i = X^i W$, where X^i is the proportion of mineral i crystallising, then

$$dW^i = X^i dW + W dX^i = -L_0 X^i dF + L_0(1 - F) dX^i \tag{4.9}$$

and Eq. (4.8) becomes

$$\frac{dc^l}{c^l} = -\frac{dF}{F} \left(1 - \sum X^i D^{i-1}\right) - \frac{1 - F}{F} \sum D^{i-l} dX^i \tag{4.10}$$

It should be borne in mind that a constraint on this equation is that

$$\sum_{i=1}^{n} X^i = 1 \quad \text{or} \quad X^i = 1 - \sum_{j=2}^{n} X^j \tag{4.11}$$

In order to make use of Eq. (4.10) it is necessary to make some assumptions about the variation of D^{i-1} and X^i. Greenland considered three possibilities, with the mineral proportion X^i as an increasing, decreasing and constant linear function of F, and showed how the trace element concentration would be affected.

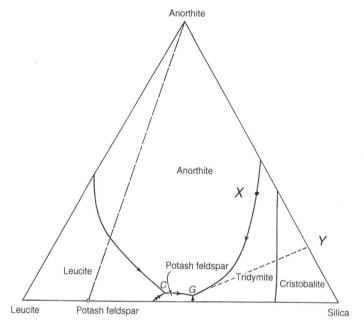

Fig. 4.1 The system leucite–anorthite–silica (after Ehlers, 1972, Fig. 40), to illustrate variation in mineral proportions crystallising from a cooling silicate melt. A melt at point X has 39% anorthite, 55% silica and 6% leucite. As it cools it precipitates anorthite and tridymite and its composition changes along the boundary curve towards point G. The proportion of anorthite–silica is given, at any point, by the intersection of the tangent at that point with the anorthite–silica join. The initial proportion is 50/50, but by the time that the liquid reaches point G, the proportion is 20/80 (point Y). Such variation affects the concentration of any trace element entering the crystallising minerals from the melt (see text).

As a similar example, Fig. 4.1 shows an example of varying mineral proportions in the anorthite–leucite–silica system; it will be assumed that the partition coefficients (mineral–melt) are constant. If a melt is crystallising to a ternary eutectic of anorthite–tridymite–potash feldspar at point G, the mineral proportions are constant and will not change, but crystallisation to a two-phase assemblage leucite–anorthite or anorthite–tridymite is quite different. In the latter case, a melt at point X has only 6% leucite and will begin to crystallise to a two-phase assemblage, where $X_{an} = X_{tr} = 0.50$.[2] As crystallisation proceeds and F decreases, the liquid composition changes along the boundary curve, whose tangent, prolonged to intersect the anorthite–silica join, gives the current proportions of the two minerals. By the time that the liquid composition reaches point G, the proportions are

---

[2] Cristobalite is the first silica phase to appear, followed, as the temperature drops, by tridymite. This will be disregarded since it may be assumed that both phases are equally inhospitable to the trace element, and $D^{cr-l} = D^{tr-l}$.

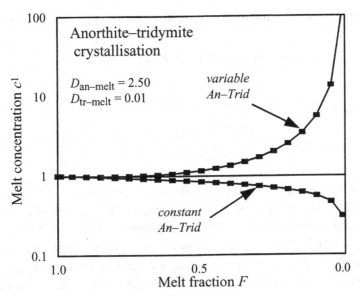

Fig. 4.2 Trace element melt concentrations during crystallisation beginning at point X in the previous figure, corresponding here to $F = 1.0$, and proceeding along the cotectic to point G. The partition coefficients are indicated, and other parameters are given in the text. One curve assumes variable mineral proportions (Eq. 4.16) as in the previous figure; after initial compatible behaviour, which is too slight to show, the element becomes incompatible and its concentration increases. The other curve shows the effect of anorthite–tridymite remaining in the initial 50/50 state throughout cooling along the boundary curve; the melt concentration decreases as the element is now behaving compatibly.

given by the point Y, where $X_{an} = 0.20$ and $X_{tr} = 0.80$. If the trace element has partition coefficients, relative to melt, of 0.01 in tridymite and 2.5 in anorthite, then the bulk coefficients (Eq. 3.8) at points X and Y are given by

$$D_X = 2.5 \times 0.5 + 0.01 \times 0.5 = 1.255 \tag{4.12}$$
$$D_Y = 2.5 \times 0.2 + 0.01 \times 0.8 = 0.508 \tag{4.13}$$

which means that element initially behaves in a compatible manner, but then becomes incompatible. In Fig. 4.2 the curve labelled *constant proportions* shows how the concentration would change if the boundary curve were straight and the initial ratio of 50/50 remained constant; the melt concentration *decreases* as crystallisation proceeds.

It will now be assumed that the proportion of anorthite is close to being a linear function of the proportion of residual melt, then

$$X_{an} = a + bF \quad \text{and} \quad X_{tr} = 1 - X_{an} \tag{4.14}$$
$$dX_{an} = b \cdot dF = -dX_{tr} \tag{4.15}$$

These quantities are substituted into Eq. (4.10), which may then be integrated to give

$$\ln \frac{c^l}{c_0} = q \cdot \ln F + r(F - 1) \tag{4.16}$$

where

$$q = (a - b)D^{an-l} + (1 - a + b)D^{tr-l} - 1 \tag{4.17}$$

$$r = 2b(D^{an-l} - D^{tr-l}) \tag{4.18}$$

The composition of melt X is 0.06 leucite + 0.39 anorthite + 0.55 silica and, as starting liquid, corresponds to $F = 1.0$; the composition of G is 0.44 leucite + 0.02 anorthite + 0.54 silica and, from the leucite proportions, $F = 0.06/0.44 = 0.136$. Then a and b are found by inserting values for the initial and final values of X_{an} and F, whence $X_{an} = 0.17 + 0.33F$. The mineral partition coefficients given above (2.5 and 0.01) are appropriate for an element such as Sr or Y, which substitutes for Ca in anorthite, but is virtually excluded by silica minerals, and Eq. (4.16) with these values is shown in Fig. 4.2, where it is seen that the melt concentration *increases* as crystallisation proceeds; the opposite of simple Rayleigh fractionation where the initial mineral proportions do not change.

Few petrologically realistic systems show two-phase boundaries with as much curvature as the leucite–anorthite–silica system, but most cotectic curves show some curvature and so it is reasonable to be concerned with the variations of dX_i in Eq. (4.10), and their effect on trace element abundances.

4.3 Variation in partition coefficients

Up to this point we have accepted that trace element partition coefficients may frequently be treated as constants. In discussing the variation of coefficients during crystallisation it is necessary to discuss trace elements and major elements separately, because partition coefficients for major elements participating in mineral solutions must always be treated as variables.

4.3.1 Trace elements

It will be now assumed that each trace element partition coefficient varies with decreasing F, and therefore with decreasing temperature. This is expected from theory (see Chapter 2) and has been used by Glitsch and Allègre (1979) to measure crystallisation temperatures using trace element data. Here it will, however, again be assumed that the dependence is linear and can be written

$$D^{i-1} = x_i + y_i F \tag{4.19}$$

This expression may now be substituted into Eq. (4.8), to obtain

$$c^l \sum (x_i + y_i F)dW_i + Ldc^l + c^l dL = 0 \tag{4.20}$$

$$\frac{dc^l}{c^l} + \frac{dF}{F} + \frac{\sum (x_i + y_i F)dW_i}{L_0 F} = 0 \tag{4.21}$$

It will also be assumed that the proportions of minerals crystallising are constant, so that $dW_i = X_i dW = -X_i L_0 dF$, whence it follows that

$$\frac{dc^l}{c^l} = \frac{dF}{F} \cdot \left(\sum X_i x_i - 1 \right) + \sum X_i y_i dF \tag{4.22}$$

and so

$$\ln \frac{c^l}{c_0} = \left(\sum X_i x_i - 1 \right) \ln F + \sum X_i y_i (F - 1) \tag{4.23}$$

and, if only one phase is crystallising

$$\ln \frac{c^l}{c_0} = (x - 1) \ln F + y(F - 1) \tag{4.24}$$

This equation is similar in structure to Eq. (4.16).

As an example, Takahashi (1978) determined the partition coefficient of olivine–silicate melt for Ni in a series of experiments under varying circumstances. In one set of experiments, with a melt of potassic basalt composition, values were 2.86 at 1450°C and 13.6 at 1200°C (Takahashi, 1978, Table 2, run numbers 24 and 34: molar values). In Fig. 4.3 suppose a completely liquid magma ($F = 1.0$) begins to crystallise to olivine at 1450°C, continuing until 60% liquid is left at 1200°C. Assuming a linear variation with F, we can use the D-values just given to calculate that

$$D^{oliv-l} = x + yF \tag{4.25}$$

where

$$x = 29.71 \text{ and } y = -26.85$$

The increase in D^{oliv-l} as crystallisation proceeds is shown in Fig. 4.3. The melt concentration decreases more quickly than a simple Rayleigh fractionation, but the concentration in the cumulate olivine rises initially until the available supply in the liquid has decreased significantly, which occurs when F equals about 90%.

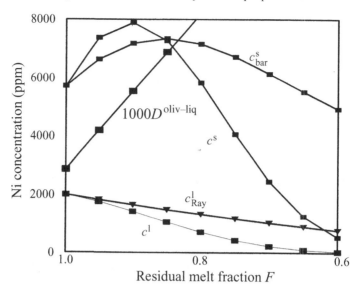

Fig. 4.3 Concentrations c^l, c^s and c^s_{bar} of Ni, during partial crystallisation of a silicate melt, with linear variation of $D^{oliv-liq}$ from 2.86 ($F = 1$) to 13.6 ($F = 0.6$); values from Takahashi (1978), see text. For comparison, the Rayleigh c^l for D constant at 2.86 is shown as c^l_{Ray}. The values of D are shown, scaled up by a factor of 1000 for convenience. The maximum in c^s occurs at $F = 0.903$.

To find the exact maximum the differential dc^s/dF is set equal to zero. Now

$$c^s = c_0(x + yF) \cdot e^v \tag{4.26}$$

where

$$v = (x - 1)\ln F + y(F - 1) \tag{4.27}$$

so

$$\frac{dc^s}{dF} = c_0 \cdot \frac{e^v}{F}(y^2F^2 + 2xyF + x(x - 1)) \tag{4.28}$$

and equating the term in brackets to zero, and solving for F gives

$$F = -\frac{2xy}{2y^2} \pm \sqrt{\frac{4x^2y^2 - 4y^2x(x - 1)}{4y^4}} \tag{4.29}$$

$$= -\frac{x}{y} \pm \sqrt{\frac{x}{y^2}} \tag{4.30}$$

With the values given above for x and y, the maximum in c^s occurs when $F = 0.903$.

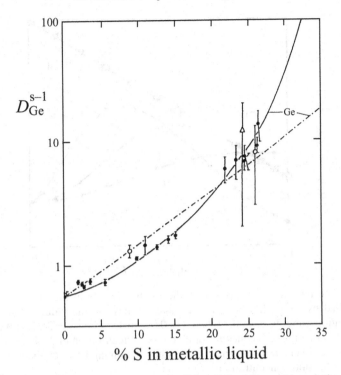

Fig. 4.4 Measurements of D_{Ge} between solid and liquid metal in relation to % S in a metallic melt consisting mainly of Fe and Ni (*solid symbols* after Jones and Drake, 1983, Fig. 1; *open symbols* earlier literature values). The coefficient is < 1 at low S contents, but rises to values an order of magnitude higher in a sulphur-rich melt.

If D remains always greater than unity then c^l shows no maximum and decreases monotonicly. Its differential is

$$\frac{dc^l}{dF} = c^l \cdot \left(\frac{x-1}{F} + y\right) \tag{4.31}$$

and is positive; this is the case in Fig. 4.3, although it does not appear so, because the scale in F is reciprocal.

It is possible for the coefficient D for mineral–melt partition to vary from an incompatible value (< 1.0) to one in excess of 1.0. Examples are also found in the partition of siderophile elements Ge, Ga and Ni between liquid metal and solid metal, in the presence of S and P, in metallic meteorites (Jones and Drake, 1983). This was discussed in Section 2.3, but another example is seen in Fig. 4.4, where D_{Ge} varies between about 0.5 with S content of 0 and 10 where $S \approx 25\%$.

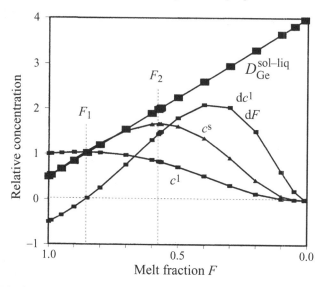

Fig. 4.5 The behaviour of Ge in metallic melts, such as in the previous figure, is modelled by $D_{Ge}^{sol-liq}$, which here is taken to have a linear dependance on S concentration, ranging from 0.5 when S = 0.0% to 4.0 when S = 25%. Initial melt ($F = 1$) has negligible S concentration and the concentration of 25% S is reached as the last liquid crystallises. There is a crystallisation lag between the peak liquid and the peak solid concentrations of Ge.

Figure 4.4 includes a line representing a linear variation of D_{Ge} over a range from 0.5 to 4.0 when S = 0 to 25%, and this is used in Fig. 4.5, assuming a melt fraction going from 1.0 to 0.0. In this case the initial values of D_{Ge} are < 1, so c^l, following Eq. (4.24), rises and reaches a maximum (where $F = F_1$). This point is difficult to see, but is obtained by setting its differential to zero, that is

$$\frac{dc^l}{dF} = c^l \cdot \left(\frac{x-1}{F_1} + y \right) = 0 \tag{4.32}$$

Now from Eq. (4.19)

$$D_{Ge} = 4.0 - 3.5F \tag{4.33}$$

and so

$$F_1 = \frac{1-x}{y} \cong 0.86 \tag{4.34}$$

The maximum in c^s occurs at a value $F = F_2$ greater than that for the maximum in c^l, ($F_2 \approx 0.57$), i.e. the peak in liquid concentration arrives before the solid concentration reaches its maximum, which Jones and Drake (1983) refer to as

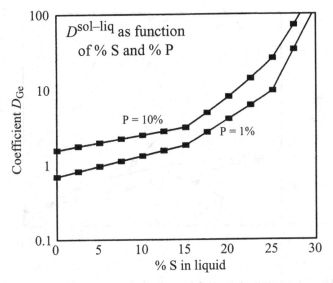

Fig. 4.6 A more realistic appraisal of the evolution of Ge concentrations in an impure metallic system than that given in the preceding Figure would use the measured values of $D_{Ge}^{sol-liq}$ from Fig. 4.4 expressed here as three functions of the concentrations of S ($< 15\%$; 15 to 25%; $> 25\%$), and two of P (1 and 10%) in equations given by Jones and Drake (1983).

a *lag*. Of course, in the case of an incompatible element the reverse will occur. The maximum in the differential of c^l occurs at a point of inflection. Jones and Drake note with reason that 'non-constant partition coefficients can produce rather non-intuitive results' (Jones and Drake, 1983, p.1203), an observation which they attribute to Kracher and Wasson (1982).

It may be overly simplistic to accept the variation of D_{Ge} with S in Fig. 4.4 as linear over the whole possible range of S concentrations. To better model the relationship, Jones and Drake divide it into three sequential linear segments and present equations which express the three regions of D_{Ge} variation shown in Fig. 4.6, under the control of yet another variable, P. Thus it is clear that, if crystallisation leads to an increase in S and P in the melt then the concentration of Ge will be affected.

4.3.2 Major elements

As already indicated, the partition coefficient of a major element in a mineral solution is always a variable. Taking the simplest case of a binary system with complete solid solution between components *i* and *j*, such as the olivines, then the solid–liquid relationships are governed by the law of mass action, as discussed

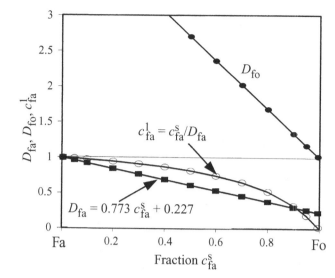

Fig. 4.7 The mass action coefficient K_D in the binary fo–fa olivine system is nearly constant, equal to 0.227, and Eq. (4.37) shows a linear variation of the solid–liquid partition coefficients D_{fo} and D_{fa} with the concentration of either component in the solid. Natural olivines conform well to this analysis.

previously, e.g.

$$K_{i-j}^{s-l} = \frac{c_i^s \cdot c_j^l}{c_j^s \cdot c_i^l} \tag{4.35}$$

where c is an atom fraction. Following Morse (1997, 2000) and many other authors we may abbreviate the exchange constant to K_D, and it is readily shown that the partition coefficients of the two components may be written as

$$D_i^{s-l} = c_i^s(1 - K_D) + K_D \tag{4.36}$$

$$D_j^{s-l} = c_j^s \left(1 - \frac{1}{K_D}\right) + \frac{1}{K_D} \tag{4.37}$$

That is, the partition coefficient of a component is a linear function of its concentration in one of the phases, provided that the exchange coefficient is a constant. This relationship was explored in detail by Morse (1997, 2000) in particular for the olivine and the plagioclase solid solution series, both for equilibrium and fractional crystallisation. The value of $K_{D, Fe–Mg}$ for the forsterite–fayalite olivines varies from 0.266 to 0.222; taking it as 0.227 then Eq. (4.37) becomes

$$D_{fa} = 0.773\, c_{fa}^s + 0.227 \tag{4.38}$$

as given in Fig. 4.7.

We have already explored the possibility that the partition coefficient for an element may have a linear dependence on the liquid fraction F during crystallisation so now we shall consider what effect a linear dependance of D_i on the concentration c_i^s will have on the liquid fraction F.

Starting with the Rayleigh equation, we examine the course of fractional crystallisation on a system such as the fo–fa olivines which follow Eq. (4.37), obtaining the following relationship (Morse, 1997, Eq. 22)

$$\frac{dF}{F} = \frac{dc^l}{ac^l(c^l - 1)} - \frac{dc^l}{c^l - 1} \tag{4.39}$$

where a ($= 1 - K_D$) is one of the constants in Eq. (4.36); this may be integrated to give

$$\ln \frac{F}{F_0} = \frac{1}{1 - K_D} \ln \left(\frac{(c^l - 1)c_0^l}{(c_0^l - 1)c^l} \right) - \ln \left(\frac{c^l - 1}{c_0^l - 1} \right) \tag{4.40}$$

In this equation F_0 equals unity, as crystallisation starts with a melt of chosen bulk composition, which therefore determines c_0^l.

For example, the starting composition in Fig. 4.8 is chosen as a melt of composition fo_{50}–fa_{50} and Eq. (4.35) is taken as

$$K_D = \frac{c_{fa}^s \cdot c_{fo}^l}{c_{fa}^l \cdot c_{fo}^s} = 0.227 \tag{4.41}$$

Calculated values of F form the abscissa. The concentrations of fo and fa vary in opposition to each other, the former decreasing to near zero while the latter – which is the lower melting component – dominates the liquid and eventually the solid phase. The fa concentration in the total solids would start at 20% rising to 50% at the end.

The use of major element partition coefficients in modelling the crystallisation of binary solutions can be extended to magmatic rock series. Langmuir and Hanson (1981) have shown how mass balance, solid-phase stoichiometry and single element partition coefficients permit the calculation of the distribution of components during crystallisation or melting, as functions of T and P. For example, the variation of $D_i^{min-melt}$ with T (already mentioned in Chapter 2) can be expressed by equations such as

$$\log_{10} D_{an}^{pl-melt} = \frac{a}{T} - b \tag{4.42}$$

where a and b are constants, for the partition of anorthite between plagioclase and melt in the system di–ab–an. Experimental measurements confirm such

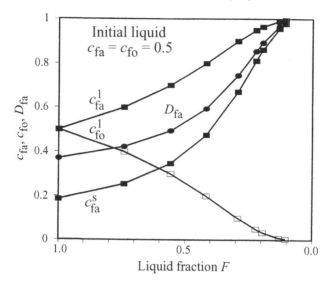

Fig. 4.8 Taking the fo_{50}–fa_{50} melt in the olivine system depicted in the preceding figure, Eq. (4.40) can be used to evaluate the evolution of fractional crystallisation. The melt concentrations c_{fo} and c_{fa} change in complementarity, since they sum to unity, the lower-melting fa increasing in the melt as crystallisation proceeds. The partition coefficient D_{fa} is linear with c_{fa}, but not with F.

relationships closely, and permit writing a *stoichiometric equation* for plagioclase as follows

$$c_{an}^l D_{an}^{pl-1} + c_{ab}^l D_{ab}^{pl-1} = 1.0 \qquad (4.43)$$

As a binary solution, the sum of the two concentrations also is unity, and the authors derive equations which permit calculation of liquid and solid compositions. In addition to binary solid solutions, examples are given of applications to a binary eutectic, a binary peritectic and more complicated systems, all related to basalt crystallisation.

These are examples of attempts to model magmatic evolution in terms of thermodynamic parameters controlled by external variables such as T and P.

Element behaviour may of course be affected by simultaneous variations of both mineral proportions and partition coefficients. If such variations are simplistically taken as linear with F in each case then model equations will be quadratic.

4.4 Crystallisation in the presence of a fluid phase

So far, the text has considered trace element behaviour in the context of a silicate melt or liquid crystallising to form cumulate minerals. The magma chamber may be replenished periodically by the parent magma, and may assimilate earlier-formed minerals or wall-rock, but these factors will be considered in the next chapter. Here the concern is the effect of separation of a fluid phase from the melt.

The melt is a multicomponent solution which usually contains compounds which are gaseous under near-surface conditions, frequently referred to as *volatile components*, and these include water, compounds of halogens, sulphur, carbon, nitrogen, etc. As minerals crystallise out of the melt, the concentrations of volatile components will increase and, if they exceed the saturation level, a separate vapour or fluid phase will appear, by a process described as *retrograde boiling, degassing, exsolution* or *vesiculation*, depending on the context. It is important to note that if the crystallisation process is governed by magma rise and decompression, then volatile loss will itself provoke extensive crystallisation (see Moore and Carmichael, 1998; Cashman and Blundy, 2000).

Volatile exsolution gives rise to many of the spectacular effects of volcanic eruptions, and may also alter the composition of surrounding wall-rocks, but here the concern lies with the influence of the chemical properties of such a solution (see Allègre *et al.*, 1977, Allègre and Minster, 1978). The system, whether open or closed, now includes two mobile multicomponent solutions – magma and fluid – and one or more solid multicomponent solutions – the minerals. The behaviour of trace element i will now be governed not only by the solid–melt partition coefficients K_i^{s-m} but also by the solid–fluid coefficients K_i^{s-f}, which are in some cases very different in magnitude.

The initial single phase melt L_0 crystallises to cumulate W_1 (Fig. 4.9a, b) either by equilibrium or fractional differentiation

$$L_0 = L_1 + W_1 \quad \text{and} \quad L_1 = FL_0 \tag{4.44}$$

At some point, F, a fluid phase W_f begins to separate (Fig. 4.9c). There are two ways to proceed. First, it may be supposed that a drastic decrease of surrounding hydrostatic pressure takes place, by intrusion of the magma body to a higher crustal level, whereupon degassing liberates essentially all of the fluid components; a second scenario will be release of fluid phase as crystallisation is proceeding.

4.4.1 Instantaneous degassing

Suppose the instantaneous release of a mass W_f leaves a residual melt mass L_2 (Fig. 4.9c). Crystallisation had produced a trace element concentration c^1 in the melt, at the point when the fluids reached a saturation concentration q and were released; the liquid concentration now became c^2.

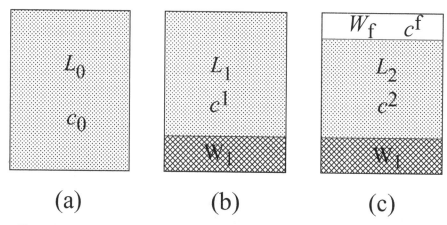

Fig. 4.9 Fractional crystallisation of a magma mass L_0 forms a cumulate W_1, leaving a residual liquid L_1 with concentration c^1. At this point the melt has become saturated in fluid components, which begin to separate with mass W_f ($= qL_1$), carrying a concentration c^f. If an instantaneous degassing takes place, then all of W_f is exsolved, leaving liquid L_2 with concentration c^2 to continue crystallising. If the ambient pressure is sufficient to prevent this, then fluid may be continually evolved as crystallisation proceeds, but maintaining the fraction q in the dwindling liquid.

So

$$L_2 = L_1 - W_f \quad \text{and} \quad W_f = qL_1 \tag{4.45}$$

$$L_2 = L_1(1-q) \tag{4.46}$$

If now we specify that the partition coefficients for the element of interest are D^{s-l} and D^{f-l}, for solid–liquid and fluid–liquid equilibria, and that the concentration in fluid is c^f, then

$$L_2 c^2 + W_f c^f = L_1 c^1 \tag{4.47}$$

$$c^f = c^2 \cdot D^{f-l} \tag{4.48}$$

If crystallisation is in equilibrium then

$$c^1 = \frac{c_0}{F(1 - D^{s-l}) + D^{s-l}} \tag{4.49}$$

and so

$$c^2 = c_0 \cdot \frac{1}{[F(1 - D^{s-l}) + D^{s-l}][1 + q(D^{f-l} - 1)]}$$

For fractional crystallisation

$$c^1 = c_0 F^{(D^{s-l}-1)} \tag{4.50}$$

and so

$$c^2 = c_0 F^{(D^{s-1}-1)} \cdot \frac{1}{1 + q(D^{f-1} - 1)} \tag{4.51}$$

These equations represent an abrupt change in concentration, but continued crystallisation of cumulates leads to further change in c^2; this will be discussed below.

4.4.2 Continued fluid release

It will be assumed that the volatile components reach saturation, as in the preceding section, but then are released continuously in proportion q to the crystallisation of cumulate minerals. The mass balance for the whole system may be written

$$L + W_s + W_f = L_0 \quad L = L_0 F \tag{4.52}$$
$$W_f = q W_s \quad dL + dW_s + dW_f = 0 \tag{4.53}$$

and so

$$dL = L_0 dF \quad dW_s = -\frac{L_0 dF}{1+q} \quad dW_f = -q \cdot \frac{L_0 dF}{1+q} \tag{4.54}$$

Two processes will again be considered – equilibrium and fractional crystallisation. In the first case the trace element mass balance may be written

$$c_0 L_0 = c^L L + c^S W_s + c^f W_f \tag{4.55}$$

and

$$c^S = D^{s-1} c^L \quad c^f = D^{f-1} c^L \tag{4.56}$$

Then, substituting and sorting terms leads to

$$c^L = \frac{c_0}{F(1-d) + d} \tag{4.57}$$

where

$$d = \frac{D^{s-1} + q D^{f-1}}{1+q} \tag{4.58}$$

It may be noted that if $q = 0$ then Eq. (4.52) becomes identical with Eq. (3.12). In the case of fractional crystallisation, the trace element mass balance is written

$$w^l + w_s + w_f = w_0 \tag{4.59}$$

and so

$$d(c^l L) + c^S dW_s + c^f dW_f = 0 \tag{4.60}$$

Substituting terms and integrating leads to the following expression

$$c^1 = c_0 F^{(d-1)} \qquad (4.61)$$

where d is the same as in the previous paragraph.

4.4.3 Discussion

Application of Eqs. (4.51), (4.57) and (4.61) is hindered by the lack of good measurements of D^{f-l} for most lithophile trace elements. There is evidence from experiment, however, that the partition is affected by anions present in the fluid phase. For example (discussed already in Chapter 3) Lagache and Duron (1987) showed that the partition of Sr between feldspar and aqueous fluid varies from 1.1 to 4.2 as the chlorinity in the fluid decreases from 2 to 0.5 (molar). Similarly Flynn and Burnham (1978) found that D^{f-l} for several of the REE increases in proportion with the cube of the chlorinity in aqueous fluid, although that for Eu is linear with the fifth power; the coefficient is, however, less than unity and the REE thus favour the melt. By contrast Keppler (1996) found coefficients for K, Rb, Pb, Zn and B are markedly increased by the presence of Cl in the fluid, and can exceed unity.

Webster *et al.* (1989) showed that for the behaviour of several cations, D^{f-l} is strongly dependent on P, T, H_2O/CO_2 and Cl. At 4 kbar and $770-950\,^\circ$C all lithophile elements have $D^{f-l} < 1$, i.e. the elements favour the melt. At lower pressures the coefficient increases for the alkalies and this effect is enhanced by decrease of temperature so that D^{f-l} for Li at 0.5 kbar and $850\,^\circ$C is 12.6. Using this value and an appropriate value for D^{wr-l} of 0.3 (Ryan and Langmuir, 1987) the equations above will now be explored.

Figure 4.10 shows the hypothetical behaviour of Li in a fractionating magma with an initial concentration of 100 ppm of Li at low P and low T, with Li coefficients as just given. The magma undergoes early crystallisation which increases the fluid content until, after 60% of the melt has crystallised, rapid degassing of 20% of the total mass takes place, carrying a high Li concentration. The c^1 for Li increases steadily from an initial concentration of 100 ppm, and then is followed by a sharp decrease to the value of c^2 (Eq. 4.40), whereupon fractionation continues. The Li-rich fluid released would probably interact with the wall-rocks enclosing the magma chamber. The value of 20% chosen for q, however, is higher than is to be expected in nature.

Continual expulsion of fluid during *the whole of a crystallisation period* is expressed using Eqs. (4.57) and (4.61); a preceding period leading to saturation in the fluid components can be added, as in Fig. 4.10. With the same partition coefficients as before, Fig. 4.11 shows the modelled concentration changes for Li during

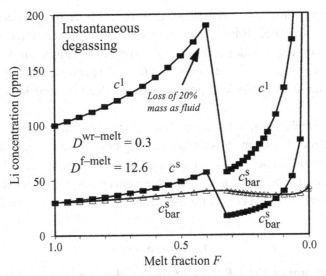

Fig. 4.10 Possible behaviour of Li in a magma undergoing fractional crystallisation. After 60% solidification ($F = 0.4$) pressure release permits instantaneous
degassing (retrograde boiling), and 20% mass ($q = 0.2$) is released as fluid, with
concentrations controlled by the D-values, of which the coefficient D^{f-l} is much
larger than D^{s-l} (Ryan and Langmuir, 1987; Webster *et al.*, 1989). The peak value
of c^l is 190 ppm, which then falls to 57 ppm (Eq. 4.40) and the concentration c^f in the
fluid released is 721 ppm (not shown). The residual melt continues to crystallise.

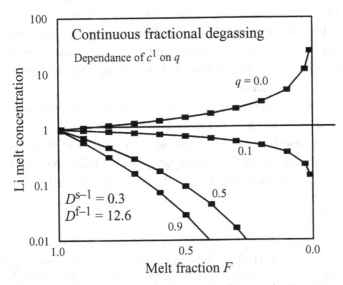

Fig. 4.11 During continuous exsolution of a fluid phase from a saturated melt
undergoing fractional crystallisation, the magnitude of q strongly affects trace
element distribution, shown here for Li, with the same coefficients as the preceding
figure. Where $q = 0.0$ there is no exsolution and Li behaves incompatibly in simple
Rayleigh fractionation, but increasing values of q lead to progressive depletion of
the melt phase. Of course a q of 0.9 is quite unrealistic.

fractional crystallisation. The melt concentration c^l is seen to depend strongly on q and on D^{f-l}. With simple Rayleigh fractionation and no fluid exsolution ($q = 0.0$) the concentration rises, under control of the slightly incompatible value of D^{s-l} (0.3). But even a small degree of exsolution ($q = 0.1$) reverses this trend, because of the high fluid–melt coefficient, and increasing q leads to drastic decrease in Li in the melt.

As a general conclusion, it has been shown that the extent to which a third phase (fluid) will affect trace element behaviour during crystallisation of a melt depends primarily on the difference between the coefficients D^{f-l} and D^{s-l}, and secondarily on the proportion of the fluid phase. Under the right circumstances the trace element will develop very different concentrations from those forming from a dry melt.

4.5 Summary

This chapter examines the influence of three additional factors on trace element fractionation during magma crystallisation.

If a magma is crystallising along a cotectic curve the mineral proportions change, and the behaviour of trace elements will be affected. Assuming constancy of partition coefficients and a variation of mineral proportions which is linear with F, changes can be made to the fractionation equation, and show that, in some cases, a trace element may change from compatible to incompatible behaviour.

When partition coefficients vary as crystallisation proceeds, a simple approach is again to assume a linear variation with F, while mineral proportions remain constant. Modifications to the fractionation equation lead to changes in the modelled concentrations, which are particularly significant for siderophile elements in metal–sulphide systems, as in meteorites.

Simultaneous variation in both mineral proportions and partition coefficients may be expected to occur, but is not easy to model.

When a fluid phase is present in addition to minerals and melt, trace element behaviour is influenced by the fluid–melt partition coefficient, but also by the mass proportions of fluid and melt, and their evolution during crystallisation.

References

Albarède, F. and Y. Bottinga (1972) Kinetic disequilibrium in trace element partitioning between phenocrysts and host lava. *Geochimica et cosmochimica acta*, **36**, 141–56.

Allègre, C. J. and J. F. Minster (1978) Quantitative models of trace element behaviour in magmatic processes. *Earth and Planetary Science Letters*, **38**, 1–25.

Allègre, C. J., M. Treuil, J. F. Minster, B. Minster and F. Albarède (1977) Systematic use of trace element in igneous process. Part I: Fractional crystallisation processes in volcanic suite. *Contributions to Mineralogy and Petrology*, **60**, 57–75.

Cashman, K. and J. Blundy (2000) Degassing and crystallisation of ascending andesite and dacite. *Philosophical Transactions of the Royal Society of London*, **A358**, 1487–513.

Ehlers, E. G. (1972) *The Interpretation of Geological Phase Diagrams*. San Francisco: W. H. Freeman.

Flynn, R. T. and C. W. Burnham (1978) An experimental determination of REE partition coefficients between a chloride-containing vapor phase and silicate melts. *Geochimica et cosmochimica acta*, **42**, 685–702.

Glitsch, L. and C. J. Allègre (1979) Determination of crystallisation temperatures in fractional crystallisation series by Ni partitioning equations. *Earth and Planetary Science Letters*, **44**, 105–18.

Greenland, L. P. (1970) An equation for trace element distribution during magmatic crystallisation. *American Mineralogist*, **55**, 455–65.

Jones, J. H. and M. J. Drake (1983) Experimental investigations of trace element fractionation in iron meteorites, II: The influence of sulfur. *Geochimica et cosmochimica acta*, **47**, 1199–210.

Keppler, H. (1996) Constraints from partitioning experiments on the composition of subduction-zone fluids. *Nature*, **380**, 237–9.

Kracher, A. and J. T. Wasson (1982) The role of S in the evolution of the parental cores of the iron meteorites. *Geochimica et cosmochimica acta*, **46**, 2419–26.

Lagache, M. and S. C. Duron (1987) Distribution of Sr between plagioclase amd 1 M aqueous chloride solutions at 600°C, 1.5 kbar and 750°C, 2 kbar. *Bullétin de la société Française de minéralogie et de cristallographie* or *European Journal of Mineralogy*, **5**, 551–62.

Langmuir, C. H. and G. N. Hanson (1981) Calculating mineral-melt equilibria with stoichiometry, mass balance and single-component distribution coefficients. In *Thermodynamics of Minerals and Melts* (eds. R. C. Newton, A. Navrotsky and B. J. Wood. (Berlin: Springer-Verlag, 1981) 247–72.

Moore, G. and I. S. E. Carmichael (1998) The hydrous phase equilibria (to 3 kbar) of an andesite and basaltic andesite from western Mexico: constraints on water content and conditions of phenocryst growth. *Contributions to Mineralogy and Petrology*, **130**, 304–19.

Morse, S. A. (1997) Binary solutions and the lever rule revisited. *Journal of Geology*, **105**, 471–82.

  (2000) Linear partitioning in binary systems. *Geochimica et cosmochimica acta*, **64**, 2309–20.

Ryan, J. G. and C. H. Langmuir (1987) The systematics of lithium abundances in young volcanic rocks. *Geochimica et cosmochimica acta*, **51**, 1727–41.

Takahashi, E. (1978) Partitioning of Ni^{2+}, Co^{2+}, Fe^{2+}, Mn^{2+} and Mg^{2+} between olivine and silicate melts: compositional dependence of partition coefficient. *Geochimica et cosmochimica acta*, **42**, 1829–44.

Webster, J. D., J. R. Holloway and R. L. Hervig (1989) Partitioning of lithophile trace elements between H_2O and H_2O-CO_2 fluids and topaz rhyolite melt. *Economic Geology*, **84**, 116–34.

5

Crystallisation assimilation, recharge and eruption

5.1 Introduction

The discussion so far has considered behaviour of a single trace element during silicate melt crystallisation as a result of fractionation and of equilibrium processes, under the influence of variation of mineral proportions and of partition coefficients and of separation of a fluid phase. In this chapter the effects of other variables will be considered, including:

(i) resorption of early minerals;
(ii) assimilation of wall-rock;
(iii) periodic recharge of the magma chamber;
(iv) loss of magma by eruption or intrusion.

5.2 Resorption or assimilation

It is possible that, during the course of magmatic crystallisation, a mineral which has crystallised in stability along a cotectic curve, becomes unstable as the temperature continues to fall. This occurs when the mineral is a product of the *incongruent melting* of a related mineral. In Fig. 5.1, enstatite melts incongruently to form forsterite olivine and melt, and this implies that, when a melt is crystallising to form olivine, then eventually the olivine becomes unstable and reacts with the melt to form enstatite. The refractory olivine undergoes *resorption*, or transformation to the liquidus phase or phases. There are many instances of this kind in silicate melt evolution (see Ehlers, 1972).

In other cases where fragments of country rock that contain less refractory minerals fall into a magma reservoir then these begin to melt and dissolve into the melt, i.e. *assimilation* takes place, the necessary heat coming from the crystallisation of a proportion p of new solids. It will be accepted in the following that *resorption* and *assimilation* are synonymous, except for the mechanisms of incorporation.

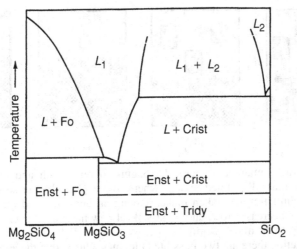

Fig. 5.1 The binary system forsterite–silica, showing the incongruent melting of enstatite $MgSiO_3$ (after Ehlers, 1972, p. 55, Fig. 41). Cooling of a liquid of the composition of enstatite, in the L_1 field, leads to crystallisation of forsterite olivine, Mg_2SiO_4; as heat is abstracted the forsterite becomes unstable and begins to react with the melt to form enstatite, i.e. it undergoes resorption. The reaction ends when all the olivine has been transformed to enstatite and all the melt is exhausted. A slight excess of Mg silicate in the initial liquid would lead to a final mixture of forsterite and enstatite. A slight excess of silica in the initial liquid would lead to complete assimilation of the forsterite before exhaustion of the melt, and a continuation of crystallisation of enstatite, joined at the eutectic by cristobalite.

There will be a transfer of any trace element in the resorbing or assimilating phase into the magma. Slow rates of diffusion of the element through solid phases may impede the process but will not be taken into account here.

5.3 Mass balance

In Fig. 5.2a a magma of mass L_0 contains a mass W_a of a mineral about to be resorbed or assimilated, and the outcome will depend in part on which is the greater. Let their ratio be Q, so

$$W_a = QL_0 \tag{5.1}$$

In cases of resorption Q will usually be < 1, but this restriction does not apply to assimilation. The previous history of the system is irrelevant so the initial liquid mass L_0 defines the residual mass fraction, F, as unity. It will be assumed that the solids W_a are resorbed in proportion P to the new minerals crystallised, leaving a residue W_r, so that

$$W_a - W_r = PW \tag{5.2}$$

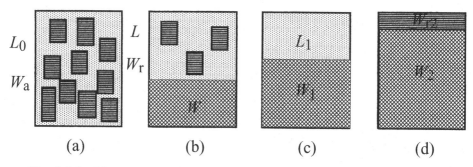

Fig. 5.2 In (a) a magma of mass L_0 begins to react with and absorb solid fragments of mass W_a (lined areas), where $W_a = QL_0$. In the second sketch (b) some assimilation has taken place, leaving an unassimilated mass W_r. The melt mass has decreased to L and a mass W of new minerals has crystallised, in proportion, P, to the mass assimilated, i.e. $W_a - W_r = PW$. If the process continues to completion there are two possible outcomes. One is that the assimilation is completed while some melt remains, shown in (c); the other is that the melt is used up, leaving some solid unassimilated (W_{r2}) as in (d). The conditions governing these possibilities are shown in the next figure. The effects of varying rates of equilibration between melt and minerals are ignored.

The system is now seen in Fig. 5.2b, where the residual liquid fraction is F, the liquid mass is L, the new solids are W and there remains W_r of the resorbing solid.

Before looking at the distribution of a trace element in such a system, it is necessary to explore the mass balance of melt and solids. Then, at this stage

$$L_0 + W_a = L + W + W_r \tag{5.3}$$

and

$$L = FL_0 \quad W_a - W_r = PW \tag{5.4}$$

so

$$L_0(1 - F) = W(1 - P) \tag{5.5}$$

and

$$W = L_0 \cdot \frac{1 - F}{1 - P} \tag{5.6}$$

Also, since $W_a = QL_0$, then $QL_0 - W_r = PW$, or

$$W_r = L_0 \left(Q - P \cdot \frac{1 - F}{1 - P} \right) \tag{5.7}$$

Substituting for W_a, L and W in Eq. (4.1), it is also readily shown that

$$F = 1 - \frac{Q}{P}(1 - P) + \frac{W_r}{L_0 P}(1 - P) \tag{5.8}$$

The stage ends *either* when all W_r has dissolved *or* all the magma has crystallised. The first is shown in Fig. 5.2c, when the liquid mass is L_1 forming a fraction F_1 where

$$F_1 = 1 - \frac{Q}{P}(1 - P) \tag{5.9}$$

and so

$$\frac{W_1}{L_0} = \frac{Q}{P} \tag{5.10}$$

If F goes to zero first, as in Fig. 5.2d, then from Eq. (5.7)

$$\frac{W_{r2}}{L_0} = Q - \frac{P}{1 - P} \tag{5.11}$$

$$\frac{W_2}{L_0} = \frac{1}{1 - P} \tag{5.12}$$

Equations (5.6) to (5.12) illustrate the effects of variation in the proportion, P, of solids assimilated. For, in the first case, when $W_r = 0$, then $F_1 > 0$, and from Eq. (5.8)

$$P > \frac{Q}{1 + Q} \qquad P_{min} = \frac{Q}{1 + Q} \tag{5.13}$$

P_{min} is the smallest value of P which is consistent with this case, leading to exhaustion of the melt at the time that the last solid dissolves (see Fig. 5.3).

In the second case, when $F = 0$ then $W_{r2} > 0$ and so

$$P < \frac{Q}{1 + Q} \qquad P_{max} = \frac{Q}{1 + Q} \tag{5.14}$$

P_{max} is the largest possible value of P, leading to simultaneous exhaustion of melt and assimilated solid, when the residual solid, W_{r2}, is zero.

The interplay of the relationships in Eqs. (5.2) to (5.12) are summarised in Fig. 5.3. The quantitative example in Fig. 5.4, starts with the choice that the mass of solid material available for resorption or assimilation amounts to one quarter of the initial magma mass i.e. $Q = 0.25$. As reaction proceeds between magma and solids the residual solids (W_r) decrease, as does the proportion of residual melt (F), according to the proportion factor P. The critical value of P, which determines whether melt outlasts solids or vice versa is P_m or $Q/1 + Q$, which here equals 0.2 and would be represented by a line from the beginning point (0.25) to the bottom-right corner, where $F = 0$. When P exceeds this value the line of evolution or reaction ends at the value of F where W_r equals zero, which for $P = 0.5$ is approximately 0.75. Where P is less than the critical value, the magma is used up ($F = 0.0$) while some

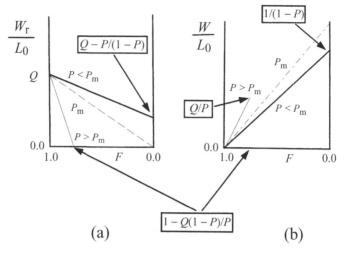

(a)   (b)

Fig. 5.3 The two possible outcomes to the assimilation process shown in Fig. 5.2 are shown here in terms of the effect of different proportions of assimilation (P) on (a) the unassimilated residue W_r, and (b) the mineral mass (W) precipitated from the melt. As discussed in the text, for a given starting value of Q, there is some value P_m such that complete assimilation occurs simultaneously with all the melt being used up, where $P_m = Q/(1 + Q)$. If P is greater than P_m (heavy line), then W_r goes to zero when $F = (1 - Q(1 - P)/P)$ and $W/L_0 = Q/P$, at which point the process stops. Alternatively, if P is less than P_m (lighter line) then all melt will have disappeared when the unassimilated residue equals $(Q - P/(1 - P))$ and the crystallised mineral mass (W_2) is $1/(1-P)$.

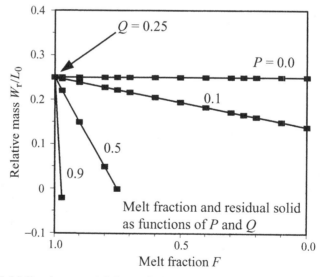

Fig. 5.4 Initially, the material for assimilation is a proportion Q of the magma mass, here taken to be 0.25. If no assimilation takes place then the proportion P is zero and the relative mass of unassimilated material (W_r/L_0) remains constant while the melt crystallises. The critical value of $P(P_m)$ is $Q/(1 + Q)$ or 0.20, and if $P = 0.1$ (i.e. less than P_m) then some assimilation takes place until the melt is exhausted, when $F = 0$. When P is equal to 0.5, i.e. greater than P_m, assimilation is complete when the remaining melt fraction is still about 0.75.

solids remain unassimilated (i.e. W_r is positive); the extreme case is where $P = 0.0$, no assimilation takes place and the magma crystallises until it is exhausted.

5.4 Assimilation by melting and solution

Having considered the solid–liquid mass balance, the trace element behaviour can now be considered. Here it is assumed that the process of assimilation proceeds by *melting, mixing and dissolution.*

The bulk mass balance (Eq. 5.2) and its differential are as follows

$$L_0 + W_a = L + W + W_r \tag{5.15}$$

$$dL + dW + dW_r = 0 \tag{5.16}$$

and, from Eqs. (5.6) and (5.7)

$$dL = dFL_0 \quad dW = -L_0 \cdot \frac{dF}{1 - P} \quad dW_r = +PL_0 \cdot \frac{dF}{1 - P} \tag{5.17}$$

Similarly for the trace element

$$w_0 + w_a = w^l + w^s + w_r \tag{5.18}$$

$$dw^l + dw_r + dw^s = 0 \tag{5.19}$$

But $w^l = c^l L$, so

$$dw^l = c^l dL + L dc^l$$
$$= L_0(c^l dF + F dc^l) \tag{5.20}$$

and

$$dw_r = c_a dW_r = L_0 \cdot \frac{c_a P dF}{1 - P} \tag{5.21}$$

The instantaneous trace element concentration in the crystals precipitated from the melt in a short period is

$$c^s = \frac{dw^s}{dW} \tag{5.22}$$

so

$$dw^s = c^s dW = -D_m c^l L_0 \frac{dF}{1 - P} \tag{5.23}$$

where c^l, c^s and c_a are the element concentrations in the melt, the crystallised solid, and the solid being assimilated; it is taken that the concentration c_a remains constant, whether the process is melting and dissolution or reaction. D_m is the bulk partition coefficient between the minerals crystallising and the melt. Substituting for the differentials in Eq. (5.19) gives the following equation

$$c^l dF + F dc^l + \frac{c_a P}{1 - P} dF - D_m c^l \frac{dF}{1 - P} = 0 \qquad (5.24)$$

This may be rearranged to

$$\frac{dc^l}{c^l(D_m + P - 1) - c_a P} = \frac{1}{1 - P} \cdot \frac{dF}{F} \qquad (5.25)$$

which, on integration gives

$$c^l = \frac{c_a P}{D_m + P - 1} + \left[c_0 - \frac{c_a P}{D_m + P - 1} \right] \cdot F^{\frac{D_m + P - 1}{1 - P}} \qquad (5.26)$$

and

$$c^s = D_m c^l \qquad (5.27)$$

The bulk trace element concentration, $\bar{c}$, in the crystallised minerals after a melt fraction F remains is given by

$$w_0 + w_a = w^l + w_r + w^s \qquad (5.28)$$

$$c_0 L_0 + c_a W_a = c^l L + c_a W_r + W \bar{c}$$

whence

$$\bar{c} = \frac{1}{1 - F} [c_0(1 - P) + c_a P(1 - F) - c^l F(1 - P)] \qquad (5.29)$$

Equation (5.26) is used in Fig. 5.5 to illustrate two cases. In the first (Fig. 5.5a), the minerals crystallising amount to ten times the mass being assimilated ($P = 0.1$). One possibility is that of a hot basic magma (basalt) assimilating a mineral (biotite) relatively enriched in an incompatible element such as Rb, for which $D_m = 0.04$, by melting and mixing. Another possibility is that the magma is assimilating a mineral enriched in a compatible element with $D_m = 15$, such as Ni. The steady addition of the Ni-rich solid to the melt quickly leads to a buffering of c^l; this end-point is marked on the graph but, of course, does not imply that in nature the crystallisation would stop; it would simply continue without assimilation, under a Rayleigh process control.

Figure 5.5b is the same as 5.5a, except that the minerals crystallising amount to twice the mass assimilated ($P = 0.5$). The behaviour of the two elements is not greatly different, but the compatible element increases its melt concentration before reaching a steady state.

The other curves on Fig. 5.5 will be examined in the next section.

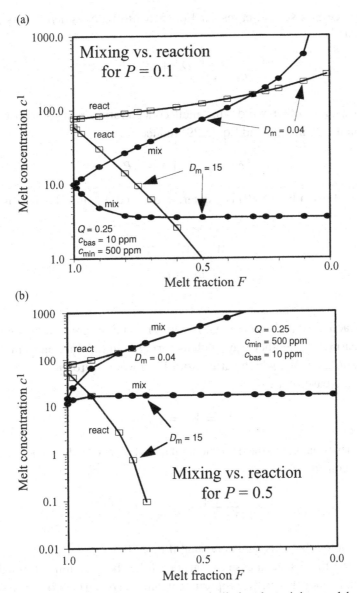

Fig. 5.5 Contrasting behaviour between assimilation by mixing and by reaction (Eqs. 5.26 and 5.51). In each case the magma mass is 4 times the material being assimilated ($Q = 0.25$), the initial trace element concentrations are the same and there are two partition coefficients, an incompatible one ($D_m = 0.04$) and a compatible one ($D_m = 15$). (a) the minerals crystallising amount to ten times the mass being assimilated ($P = 0.1$) in this case crystallisation continues until all the liquid has gone, leaving an incomplete assimilation (see Fig. 5.2d). (b) the minerals crystallising amount to twice the mass being assimilated ($P = 0.5$). In this case, and if the assimilation is by reaction, then all the solids will have gone when the residual magma proportion is ≈ 0.70.

There are certain special cases for Eq. (5.26) to be considered. First, when it happens that

$$c_a = c_0 \cdot \frac{D_m + P - 1}{P} \tag{5.30}$$

then $c^l = c_0$, or in other words, the melt concentration remains constant while assimilation is continuing, and Eq. (5.29) reduces to

$$\bar{c} = c_0(1 - F) + c_a P \tag{5.31}$$

Next, if $P = 1$, then Eq. (5.19) is indeterminate, and the mass balance is changed as follows

$$W_a - W_r = W \tag{5.32}$$
$$L_0 = L \tag{5.33}$$
$$dW_r = -dW \tag{5.34}$$

So F, the fraction of melt remaining, is constant, and progress in the assimilation must be measured in another way, such as variation in the fraction of material left for assimilation, i.e. W_r, which varies from W_a down to zero, *or* W, which begins at zero and increases until

$$W = W_a = QL_0 \tag{5.35}$$

Following these changes to the trace element mass balance leads to a new expression for the melt concentration

$$c^l = \frac{c_a}{D_m} + \left[c_0 - \frac{c_a}{D_m} \right] e^{-\frac{QD_m W}{L_0}} \tag{5.36}$$

It should be noted that c^l is again constant if $c_a = Dc_0$.

There is one more special case to consider. It has been assumed throughout the preceding discussion that P, the proportion of material being resorbed or assimilated will be less than unity, i.e. less than the proportion of minerals being crystallised. This is undoubtedly true in most instances where excess heat is not available. But in the case of hot ultramafic magma assimilating low-melting feldspar or mica the situation may be different, with $P > 1$. The result is that when the assimilation is complete, i.e. when $W_r = 0$ and $L = L_1$ (see Fig. 5.2), then since

$$P = \frac{W_a - W_r}{W} \tag{5.37}$$

in consequence

$$L_0 + W_a = L_1 + W \tag{5.38}$$

$$L_1 - L_0 = W_a - W = (P - 1)W \tag{5.39}$$

Since $P > 1$, thus $L_1 > L_0$ and also $F > 1$. The increase in the melt mass must be fuelled by excess melt heat.

5.5 Assimilation by reaction

The previous section treats the process of resorption or assimilation as one of melting and mixing, in which the trace element concentration c_a does not change, and consequently its partition coefficient D_a does not appear. This is not realistic if more refractory material is being resorbed, when reaction will accompany dissolution.

Analysis of this situation is not straightforward and requires a preliminary assumption that the trace element equilibrates between the melt and the solid before crystallisation starts, to give concentrations c_0^1 and c_a^0, respectively:

$$c_0^1 = \frac{c_0 + Q c_a}{1 + Q D_a} \quad c_a^0 = D_a c_0^1 \tag{5.40}$$

After this stage, however, the analysis may proceed under the assumption that the residual concentration in the resorbate will now be controlled by that in the melt, i.e.

$$c_r = D_a c^1 \tag{5.41}$$

It is assumed that the material is a mineral assemblage such that

$$D_a = \sum_i D^{i-m} r_i \tag{5.42}$$

where r_i is the proportion of mineral i. Then the trace element mass balance is

$$w_0 + w_a = w^1 + w^s + w_r \tag{5.43}$$

so

$$dw^1 + dw_r + dw^s = 0 \tag{5.44}$$

and

$$dw^1 = d(c^1 L) \quad dw_r = d(c_r W_r) \tag{5.45}$$

As in the previous section we define the instantaneous concentration in the precipitated solids as

$$c^s = \frac{dw^s}{dW}$$

(5.46)

and also we have

$$c_r = D_a c^l \quad c^s = D_m c^l$$

(5.47)

where D_m is the bulk coefficient for the minerals crystallising from the melt, D_a is the bulk coefficient for the material being resorbed, and c_r is the trace element concentration in that material. So

$$c^l dL + L dc^l + c_r dW_r + W_r dc_r + c^s dW = 0$$

(5.48)

As before

$$dW = -dF \cdot \frac{L_0}{1-P} \quad dW_r = dF \cdot \frac{L_0 P}{1-P}$$

(5.49)

and, of course

$$dL = L_0 \cdot dF$$

(5.50)

so by rearrangement and integration the variation of the trace element is given by

$$c^l = c_0^l \cdot \left(\frac{x + yF}{x + y} \right)^{\frac{z}{y}}$$

(5.51)

and, from above

$$c_r = D_a c^l$$

(5.52)

where

$$x = D_a(P + PQ - Q)$$

(5.53)

$$y = P - PD_a - 1$$

(5.54)

$$z = 1 - P - D_m + PD_a$$

(5.55)

It is to be noted that if $Q = 0$ and $P = 0$, Eq. (5.36) reduces to a simple Rayleigh fractionation (Eq. 3.20).

As in the last section, the bulk trace element concentration, $\bar{c}$, in the crystallised minerals after a melt fraction F remains is given by

$$w_0 + w_a = w^l + w_r + w^s$$

(5.56)

$$c_0^l L_0 + c_a^0 Q L_0 = c^l L + c_r W_r + W\bar{c}$$

(5.57)

whence

$$\bar{c} = \frac{\left(c_0^1 + Qc_a^0\right)(1 - P)}{1 - F}$$

$$- \frac{c^1}{1 - F}\left(D_a(Q - QP - P) + F(1 - P + PD_a)\right) \qquad (5.58)$$

The results of this kind of assimilation process, using Eq. (5.51), are shown in Fig. 5.5a and 5.5b, which have been constructed in the same way as the last section, except for the assumption here of equilibration between the assimilating solid and the melt, before the assimilation takes place. If the mineral being assimilated is rich in the trace element, as is the case with both elements considered here, then this equilibration results in a strong enrichment in the melt, whether the element is incompatible or compatible. In Fig. 5.5a, where P is small, the incompatible element concentration rises slowly in the melt, whereas the incompatible element decreases steeply. But in Fig. 5.5b, where P is 0.5, the solid is rapidly absorbed and disappears by the time that the melt fraction reaches about 70%, as illustrated in Fig. 5.2c.

5.6 Assimilation-fractional crystallisation processes

The effects of assimilation on trace element concentrations and isotopic ratios, while a silicate melt is undergoing crystallisation, were examined in detail by DePaolo (1981). The *'assimilation-fractional crystallisation'* (AFC) model which he constructed includes the theory presented in Section 5.2, but also includes an open system option where the magma chamber may be replenished from time to time. DePaolo's Eq. (6a), which expresses the variation of trace element concentration, is identical to Eq. (5.26) above. AFC theory will be examined again in Section 5.8, after first modelling volcanic systems which can erupt and can be recharged.

5.7 Magma recharge and discharge

The previous sections have considered the effects on trace element concentration of relaxation of the closed system restriction treated in Chapter 3, namely the introduction of solid rock or mineral to the magma chamber. Next it is appropriate to discuss the effects of further relaxation by addition of new magma and the loss of magma by eruption or intrusion, thereby approaching more closely natural igneous systems.

Although repeated eruptions at volcanic centres had long suggested that the reservoirs which fed them must be replenished from time to time, attempts to construct useful quantitative models, for application to trace element behaviour, were pioneered by Michael O'Hara (1977; O'Hara and Matthews, 1981) and

Francis Albarède (1985, 1995). Those works exploit the idea of a magma chamber or reservoir within the Earth's crust, which is cooling and undergoing fractional crystallisation, while wall-rocks are subjected to assimilation, periodic eruptions of lava or intrusions into the adjacent crustal rocks occur and replenishment by a new supply of the (parent) magma takes place from time to time. These processes may or may not lead to a steady-state situation. Further development of theory has been provided by Pankhurst (1977), Caroff (1997) and Caroff *et al.* (1997).

Strong support for these processes comes from layered ultramafic complexes, where repeated sequences may occur. One such is the Muskox intrusion in north-west Canada, where four episodes of magma recharge (see Fig. 5.6) have been identified (Irvine and Smith, 1967). Another is on the isle of Mull, where 20 pulses of magma recharge have been documented by Usselman and Hodge (1978) who proposed a temporal and thermal history of this complex; an oscillation of the liquidus temperature around $1220\,^\circ$C led to change in the crystallising assemblage, and consequent alternation of the rock types dunite and allivalite.

With such periodic processes a number of simplifying assumptions are necessary to construct a realistic model, and careful definitions of the events and their sequencing are necessary. For instance, it is convenient to depict the eruption, replenishment and assimilation events as proportional to the size of the reservoir, but it is then important to specify whether it is the *original* or the *residual* size: O'Hara only referred to 'the steady state mass, M of the magma chamber' (1977, p. 505). Also, the events considered may influence the *volume* of the chamber, so this must be considered.

Recent developments in geochronology give some insight into the actual time sequences of events at real volcanic centres (see Hawkesworth *et al.*, 2000).

5.7.1 Conservation of initial magma mass

Beneath a volcanic edifice the magma chamber, depicted in Fig. 5.7, undergoes fractional crystallisation. Periodically the volcano erupts and magma is extruded; periodically new parent magma is injected into the magma chamber; the walls of the chamber continually undergo alteration and assimilation by the magma. These processes, and the sequence in which they occur, will influence the behaviour of trace elements in the magma and the cumulates. It will be assumed that the evolution is cyclical, and abbreviations will be

C+A   crystallisation with assimilation
E   eruption
R   recharge

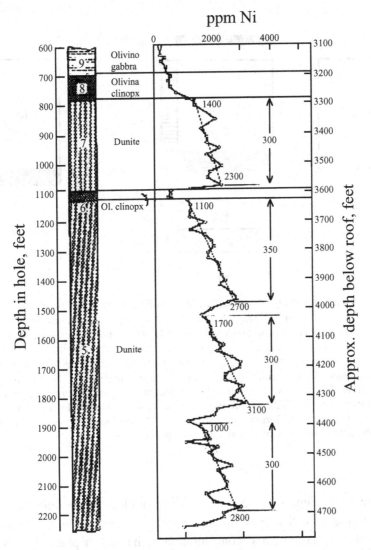

Fig. 5.6 In the Muskox complex, Northwest Territories, Canada, Irvine and Smith (1967, Fig. 2.7) show that the correlation with depth of Ni content of samples taken by deep drilling reveals the presence of four distinct sequences of dunite, representing periodic recharge of a magma reservoir, followed each time by fractional crystallisation.

The chamber is initially filled by a mass L_0 of parent magma, which differentiates to yield a liquid mass L_1 and solid cumulate mass W_1 (Fig. 5.7a,b), where $L_1 = F_1 L_0$ and $W_1 = (1 - F_1) L_0$; the trace element has concentrations c_0, c^1, c^s and c_a in the parent magma, the differentiated magma, the solids being precipitated and the rocks being assimilated.

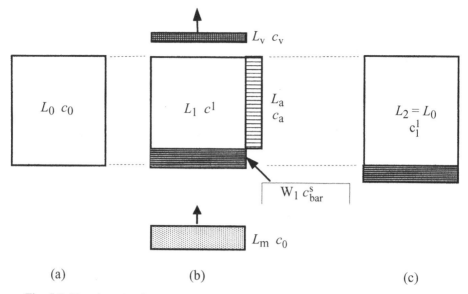

Fig. 5.7 Fractionation in a magma chamber with periodic eruption, recharge and wall-rock assimilation. (a) Magma mass L_0 with trace element concentration c_0. (b) Fractional crystallisation produces a mass W_1 of cumulate minerals, with bulk concentration c^s_{bar}, leaving a liquid concentration of c^1, while assimilating a mass L_a of wall-rock with concentration c_a; a mass L_v is next erupted as lava or injected as an intrusion; its concentration c_v is, in this case, equal to c^1. Then the magma chamber is recharged with a mass L_m of parent magma, carrying the original concentration c_0. (c) The residual magma is the mass L_2, identical with the starting mass L_0, but with a different concentration c^1_1. Repetition of the cycle may now take place, leading perhaps to a steady-state process.

$C + A \rightarrow E \rightarrow R$

Next, there is an eruption of mass L_v with trace element concentration c_v, followed by a magma recharge of L_m at the parent concentration, c_0. Assimilation of wall-rocks of mass L_a and concentration c_a has taken place. It will be assumed that the liquid mass L_0 is conserved as changes take place, and during this stage it

    (i) loses a proportion x as cumulate;
   (ii) loses a proportion y as lava;
  (iii) gains a proportion z as new magma.

The liquid may also gain a proportion Px by assimilation, assuming as in Section 5.3, that the mass assimilated is proportional to the cumulate mass crystallised. It should be noted that the duration of such assimilation will coincide with the crystallisation.

The assumption of constancy of liquid mass L_0 requires that

$$(L_0 - L_1) + L_v = L_m + L_a \tag{5.59}$$

So

$$x L_0 + y L_0 = z L_0 + P x L_0$$

or

$$z = x + y - Px \tag{5.60}$$

Recalling that $x = 1 - F_1$, then

$$z = y + (1 - F_1)(1 - P) \tag{5.61}$$

The mass balance for the trace element in the liquid, at the end of this stage, can be written

$$c_1^l L_0 = c^l L_1 - c_v L_v + c_0 L_m + c_a L_a \tag{5.62}$$

using c^l for the magma concentration after fractionation, where

$$c^l = c_0 F_1^{D-1} \tag{5.63}$$

The instantaneous concentration in the solids crystallising is $c^s = Dc^l$, but the accumulated concentration is

$$\bar{c}^s = \frac{1 - F_1^D}{1 - F_1} \tag{5.64}$$

Now Eq. (5.62) can be reduced to

$$c_1^l = F_1 c_0 F_1^{D-1} - c_v y + c_0 z + c_a P(1 - F_1) \tag{5.65}$$

In the present case, where eruption *follows* the fractional crystallisation, the concentration c_v is equal to c^l, and so

$$c_1^l = c_0(F_1 - y)F_1^{D-1} + c_0 z + c_a P(1 - F_1) \tag{5.66}$$

where

$$z = y + (1 - F_1)(1 - P) \tag{5.67}$$

In the case where assimilation is disregarded Eq. (5.66) becomes

$$c_1^l = c_0 F_1^{D-1}(F_1 - y) + c_0(y + 1 - F_1) \tag{5.68}$$

This is the end of the first cycle, and the parameters for the beginning of the second are L_0, F_2 and c_1^l. After fractionation, the mass balance and liquid concentration are

$$L_0 = L_2 + W_2 \quad L_2 = F_2 L_0 \tag{5.69}$$

$$c_2^l = c_1^l F_2^{D-1} \tag{5.70}$$

so, substituting into Eq. (5.66) gives

$$c_2^l = c_1^l F_2^{D-1}(F_2 - y) + c_0\,(y + (1 - F_2)(1 - P)) + c_a P(1 - F_2)$$

and for ensuing steps

$$c_n^l = c_{n-1}^l(F_n - y)F_n^{D-1} + c_0\,(y + (1 - F_n)(1 - P)) + c_a P(1 - F_n) \tag{5.71}$$

If the value of F remains constant, Eq. (5.71) leads to a steady-state evolution, and the ultimate liquid concentration (for a given value of $D > 1$) depends on $(F - y)$; when positive $c_n^l < c_0$, when negative c_n^l increases steadily, and when zero there is a steady state where c_n^l remains constant equal to c_0. Of course if $D < 0$ these relations are reversed.

An example of the concentration patterns generated by Eq. (5.71) is shown in Fig. 5.8a, which shows the behaviour of Ni during three cycles of (C+A) $\rightarrow$ E $\rightarrow$ R for a basic magma containing 100 ppm Ni crystallising to a dunitic mineral assemblage (no assimilation). With a partition coefficient of 10, the initial solids contain 1000 ppm Ni, then the concentration in the liquid decreases, and reaches a steady state when the amount of Ni separating out in the cumulate equals the amount brought in by the recharge. For example, with the parameters shown, the Ni concentration in the melt decreases to 7.508 ppm in the first cycle and, since $F = 0.75$, $x = 1 - F = 0.25$ and $y = 0.50$, then by Eq. (5.68)

$$c_1^l = 7.508(0.75 - 0.50) + 100(1 - 0.75 + 0.5) \tag{5.72}$$

$$= 76.88$$

The concentration in the solid separating is 769 ppm, and the mass of Ni in the first cumulate is

$$(1 - F)\bar{c}^s = 100(1 - 0.75^{10}) = 94 \tag{5.73}$$

whereas the mass brought in is

$$(1 - F + y)c_0 = 100(1 - 0.75 + 0.5) = 75 \tag{5.74}$$

The system has almost settled to a steady state in the second cycle, when the mass of the cumulate becomes nearly equal to the mass of the recharging magma. The sketch in Fig. 5.8b shows a possible Ni distribution in dunitic layers upwards from the magma chamber floor, i.e. in a similar situation to Fig. 5.7a and b.

Incompatible elements do not reach a steady-state concentration, as will be discussed below.

(a)

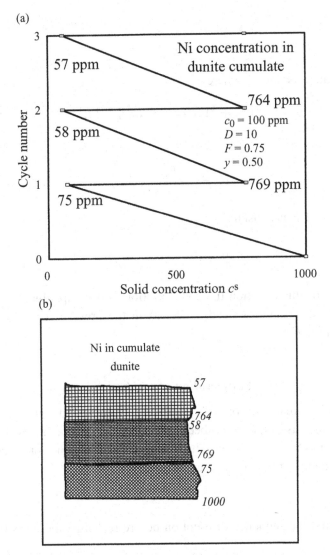

(b)

Fig. 5.8 Nickel concentrations in a magma chamber cumulate, where the partition coefficient $D^{Ni-melt} = 10$. The initial magma, containing 100 ppm Ni, undergoes 25% fractional crystallisation, followed by discharge of 50% of the liquid mass in an eruption, then is recharged with fresh parent magma, as in Fig. 5.7: (a) the instantaneous Ni concentrations in the dunite minerals, and (b) a sketch of the cumulates on the magma chamber floor, similar to the Muskox complex (Fig. 5.6).

$C+A \rightarrow R \rightarrow E$

When recharge precedes eruption, then slight changes are needed to the equations in the previous section. After crystallisation the trace element concentration is c^1, as before, then a recharge takes place, which will bring the magma mass to $FL_0 + zL_0$; this exceeds L_0 but after eruption the mass is again conserved and so, as before

$$z = 1 - F + y \tag{5.75}$$

The concentration is now c_1^1 where

$$c_1^1(F_1 + z)L_0 = c^1 F_1 L_0 + c_0 z L_0 \tag{5.76}$$

whence

$$c_1^1 = c_0 \frac{F_1^D}{1 + y} + c_0 z \tag{5.77}$$

and, in general, if F is constant

$$c_n^1 = c_{n-1}^1 \frac{F^D}{1 + y} + c_0 z \tag{5.78}$$

Under control of this equation the concentration of a compatible element in the liquid quickly settles into a steady state, as in the previous case, whatever the values chosen for F and y.

5.7.2 Conservation of residual magma

In some published treatments of the crystallising, erupting and replenishing magma chamber the basic assumption has been taken that the residual magma mass after crystallisation has been conserved. The following considers the consequences, following a treatment parallel to the previous sections.

C+A → E → R

Considering first the sequence of eruption before recharge, the mass balance for the magma is

$$L = L_1 - L_v + L_m + L_a \tag{5.79}$$

where L is the magma mass at the end of the cycle, and the other symbols are as defined previously. But now it is the residual magma mass L_1 which is conserved, so we have

$$L = F_1 L_0 - y L_1 + z L_1 + P x L_1 \tag{5.80}$$

and the condition for conservation is that $L = L_1 = F_1 L_0$, so that

$$z = y - P(1 - F) \tag{5.81}$$

Next, disregarding the assimilation term, the mass balance for the trace element can be written

$$c_1^l L_1 = c^l L_1 - y c^l L_1 + z c_0 L_1 \tag{5.82}$$

where

$$c^l = c_0 F_1^{D-1} \tag{5.83}$$

so

$$c_1^l = c_0 \left((1-y) F_1^{D-1} + y \right) \tag{5.84}$$

In the second cycle the concentration is given by

$$c_2^l = c_1^l (1-y) F_2^{D-1} + c_0 y \tag{5.85}$$

and, similarly for the general case

$$c_n^l = c_{n-1}^l (1-y) F_n^{D-1} + c_0 y \tag{5.86}$$

C+A → R → E

A similar reasoning for a process of crystallisation followed by recharge, then eruption (C+A → R → E), leads to the following mass balance

$$L^* = L_1 + L_{\mathrm{m}}$$
$$= L_1(1+z) \tag{5.87}$$

The liquid concentration is now c_1^l and so

$$c_1^l L^* = L_1 c_0 F_1^{D-1} + z L_1 c_0 \tag{5.88}$$

Next the eruption of mass $y L_1$ returns the liquid mass to its steady-state value L_1 and so

$$L_1 = L^* - y L_1 \tag{5.89}$$

and so again z = y.
Substituting this in the equation for c_1^l and generalising, leads to the following

$$c_n^l = c_{n-1}^l \cdot \frac{F_n^{D-1}}{1+y} + c_0 \cdot \frac{y}{1+y} \tag{5.90}$$

5.7.3 Discussion

Figure 5.9a and b shows the melt concentration c_n^l in a few cycles of a fractionating–recharging and erupting magma system, which disregards assimilation processes

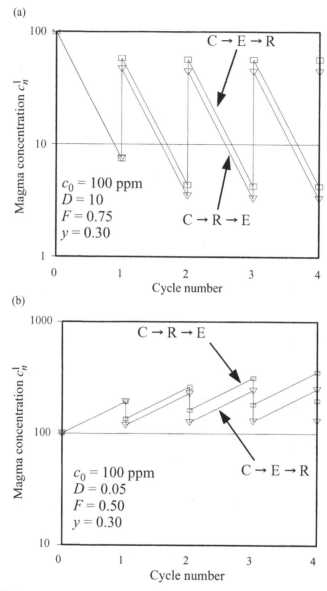

Fig. 5.9 Melt concentrations obtained from the two processes labelled C+A → E → R and C+A → R → E in Section 5.2.1 of the text and expressed by Eqs. (5.71), (5.78) for (a) a compatible and (b) an incompatible element. There is marked similarity between these two models, i.e. it does not much matter whether eruption precedes or is subsequent to recharge. Concentrations are higher for the compatible element when eruption precedes recharge; it is the reverse for the incompatible element. A steady state is achieved after four or five cycles, at a concentration in the vicinity of c_0/D.

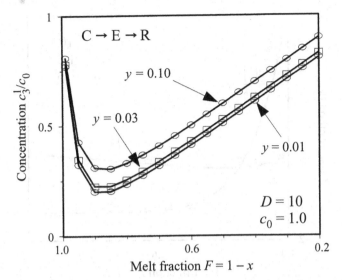

Fig. 5.10 The structure of Fig. 5.9 is a function of the parameters D, F and y. The variation of c_n^l is non-linear and reaches a minimum when $D > 1$. The magnitude of c_n^l depends on y, but the F-value of the minimum is virtually constant.

but assumes that the magma mass L_0 remains constant at the end of each cycle. The concentrations have been calculated from Eqs. (5.71) and (5.78) for a compatible and an incompatible element; the calculations have been made for the case of eruption preceding recharge and of the reverse. In each case the eruption amounts to 30% of the melt mass.

The first point to note is the overall similarity of the two models, $C \rightarrow E \rightarrow R$ and $C \rightarrow R \rightarrow E$, i.e. it does not much matter whether eruption precedes or is subsequent to recharge. When eruption comes first, the concentrations are slightly higher for the compatible element, and slightly lower for the incompatible element, relative to when recharge comes first. Secondly, a steady state is achieved after four or five cycles, at a concentration in the vicinity of c_0/D.

Similar calculations for Eqs. (5.86) and (5.90) show that conservation of the original magma mass gives higher concentrations than conservation of the residual mass for the incompatible element (and the reverse for the compatible one), but the patterns are very similar; this is to be expected and the concentration curves are displaced by an amount close to the element loss in the first cycle.

Of course in Fig. 5.9 arbitrary choices were made for the parameters D, F and y. The effects of variations, although maintaining the constraint that y can not exceed F, are seen in Fig. 5.10. For the compatible element shown in this case, the melt concentration in the third cycle increases as the proportion of lava erupted (i.e. the value of y) increases; but the concentration also depends on the melt fraction

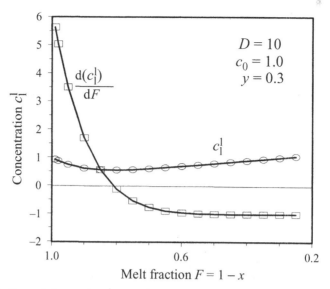

Fig. 5.11 The minimum in c_n^l seen in Fig. 5.10 occurs at a value of F where $dc_n^l/dF = 0$ and can be calculated from that equality.

after crystallisation, reaching a pronounced minimum at about $F = 0.87$. These dependencies are further illustrated in Fig. 5.11 which shows the differential of Eq. (5.68) and the melt concentration in the first cycle. The differential is given by

$$\frac{dc_1^l}{dF} = c_0 F^{D-2}[DF - y(D-1)] - c_0 \tag{5.91}$$

and this equation equals zero at the minimum. The minimum is at $F \approx 0.8$ from Fig. 5.11, and is calculated to be $F = 0.810$.

   The evidence from the preceding paragraphs examines the models of fractionating, erupting and recharging magma chambers described in the references given at the beginning of this section. Many recent publications have amplified those references (e.g. O'Hara, 1995; O'Hara and Fry, 1996) but most authors still appear to accept that trace element concentrations show steady-state behaviour, although it is clear that such claims are too sweeping.

5.8 Recharge, eruption, assimilation: the rate process model

The discussion in the preceding sections has modelled magma chamber evolution as a function of the magma fraction F. Although very useful, this is a rather abstract concept, which cannot be measured in nature and some authors believe that more realistic variables are the rates at which various reactions proceed. The conceptual relationship between the two systems has already been mentioned in Section 3.3.

This approach was introduced by DePaolo (1981), as mentioned in Section 5.6, and, because it has been adopted by many workers under the name of AFC theory it seems desirable to review the formalism and some of the applications at this point. First we must define the following:

rate of assimilation	$dM_a/dt = R_a$
rate of magma recharge	$dM_i/dt = R_i$
rate of crystallisation to solids	$dM_s/dt = R_s$
rate of eruption	$dM_e/dt = R_e$
assimilant concentration	c^a
input magma concentration	c^i take equal to initial value c_0
crystallised solids concentration	c^s
current magma concentration	c^m
erupted lava concentration	c^e take equal to current magma c^m, whether before or after mixing

M = magma mass
m = trace element mass in magma
$D = c^s/c^m$

We start with an equation for the variation of m during crystallisation of the magma

$$\frac{dm}{dt} = d(c^m M) = M\frac{dc^m}{dt} + c^m\frac{dM}{dt} \tag{5.91}$$

If assimilation is taking place then we can also write

$$\frac{dm}{dt} = R_a c^a - R_s c^s = R_a c^a - R_s c^m D \tag{5.92}$$

and

$$\frac{dM}{dt} = R_a - R_s \tag{5.93}$$

Rearrangement gives DePaolo's Eq. (1b), for concentration behaviour during crystallisation and assimilation, i.e.

$$\frac{dc^m}{dt} = \frac{R_a}{M}(c^a - c^m) - \frac{R_s}{M}(D - 1)c^m \tag{5.94}$$

We can add terms for replenishment or recharge of magma and eruption and write that

$$\frac{dm}{dt} = R_a c^a + R_i c^i - R_s c^s - R_e c^e \tag{5.95}$$

So, from (5.90) and (5.94)

$$\frac{dc^m}{dt} = \frac{1}{M}\left(R_a c^a + R_i c^i - R_s c^s - R_e c^e\right) - \frac{c^m}{M}\cdot\frac{dM}{dt} \qquad (5.96)$$

also we know that

$$\frac{dM}{dt} = R_a + R_i - R_s - R_e \qquad (5.97)$$

Whence, assuming that $c^e = c^m$ we get

$$\frac{dc^m}{dt} = \frac{R_a(c^a - c^m) + R_i(c^i - c^m) - R_s c^m(D-1)}{M} \qquad (5.98)$$

5.8.1 Conservation of the initial magma mass

We now assume that the magma chamber is of constant size (volume, mass) although the space occupied by the cumulates is ignored, as in Fig. 5.7. In other words, the ratio of initial to final magma mass (F) is 1, then

$$\frac{dM}{dt} = 0 \quad \text{or} \quad R_a + R_i - R_s - R_e = 0 \qquad (5.99)$$

and, putting $c^i = c_0$, Eq. (5.96) becomes

$$\frac{dc^m}{dt} = \frac{R_a c^a + R_i c_0}{M} - c^m \cdot \frac{R_s D + R_e}{M} \qquad (5.100)$$

So

$$\frac{dc^m}{dt} = p - q c^m \qquad (5.101)$$

where

$$p = \frac{R_a c^a + R_i c^i}{M} \qquad q = \frac{R_s D + R_e}{M} \qquad (5.102)$$

The integral is

$$\frac{p - q c^m}{p - q c_0} = e^{-qt} \qquad (5.103)$$

or

$$c^m = c_0 e^{-qt} + \frac{p}{q}(1 - e^{-qt}) \qquad (5.104)$$

It is to be noted that

$$\frac{p}{q} = \frac{r_a c^a + r_i c_0}{D + r_e} \qquad (5.105)$$

where

$$r_a = \frac{R_a}{R_s} \quad r_i = \frac{R_i}{R_s} \quad r_e = \frac{R_e}{R_s} \tag{5.106}$$

The mass of solids formed during time t is $M_s = R_s t = M(1 - F)$, so

$$qt = \frac{R_s D + R_e}{M} \cdot t$$
$$= (r_e + D)(1 - F) \tag{5.107}$$

The formulations in Eqs. (5.104) and (5.107) are now in *relative*, rather than *absolute* rates, which are in most cases easier to use. It should also be noted that, in terms of the definitions in Section 5.7 that

$$r_e = y \quad r_i = z \quad r_a = Px \tag{5.108}$$

5.8.2 Magma mass M is not constrained

In this case F is variable and it is convenient to convert time t to F, as detailed in Section 3.3: a relative change in magma mass is dM/M, and if this is integrated from M_0 to M we get $\ln M/M_0 = \ln F$, so that

$$d \ln F = \frac{dM}{M} \tag{5.109}$$

and this can now be substituted into Eq. (5.97) to obtain

$$\frac{d \ln F}{dt} = \frac{R_a - R_s + R_i - R_e}{M} \tag{5.110}$$

or

$$d \ln F = \frac{R_s dt}{M}(r_a + r_i - r_e - 1) \tag{5.111}$$

The melt concentration, from Eqs. (5.96) and (5.97) is

$$\frac{dc^m}{dt} = \frac{R_a c^a + R_i c^i - R_s c^s - R_e c^e}{M} - \frac{c^m}{M}(R_a + R_i - R_s - R_e)$$
$$= \frac{R_s}{M}(r_a c^a + r_i c^i) - \frac{R_s c^m}{M}(r_a + r_i + D - 1) \tag{5.112}$$

and so from Eq. (5.109)

$$dc^m = d\ln F \cdot \frac{r_a c^a + r_i c^i - c^m(D - 1 + r_a + r_i)}{r_a + r_i - r_e - 1} \tag{5.113}$$

which may be integrated to give

$$c^m = c_0 F^{-q} + \frac{p}{q}(1 - F^{-q}) \tag{5.114}$$

where

$$p = \frac{r_a c^a + r_i c^i}{r_a + r_i - r_e - 1} \quad \text{and} \quad q = \frac{D - 1 + r_a + r_i}{r_a + r_i - r_e - 1} \tag{5.115}$$

It should be remembered that c^m is also equal to c^e.

A number of authors have used the approach just outlined. Reagan *et al.* (1987) interpreted the nearly continual eruption of the Arenal volcano in Costa Rica in terms of magma fractionation followed by intrusion of new magma which mixed with previous mafic fractions, which they describe as *open system differentiation*. Their modelling follows a similar sequence to the present one, looking first at a constant magma mass (with their Eqs. (1), (2) and (7) the same as (5.96), (5.97) and (5.104)[1]) and then at a system with unconstrained magma mass (their Eqs. (8) and (9) or (5.111) and (5.114)).

Caroff *et al.* (1997) use analyses of erupted lavas from the East Pacific Rise to develop a model of a cyclically inflating and deflating magma chamber of changing volume, or *non-steady state*. Their approach starts from Reagan *et al.*'s Eq. (9), expressed here as Eq. (5.114), and takes it as their Eq. (1).[2] The character of a deflating chamber is that the mass of liquid in the chamber is decreasing, and the authors use the measure Δ, where

$$\Delta = r_i - r_e - 1 \tag{5.116}$$

When $\Delta > 1$ the chamber is inflating and the melt fraction, F, will move to values exceeding unity. The behaviour of an incompatible element in both an inflating and a deflating chamber is shown in Fig. 5.12, and the trends are different. For comparison, the concentration in a chamber of fixed capacity is included. To facilitate such comparisons the authors also use two other variables; one is the magma flux, measured by r_i and r_e; the other is a dimensionless time measure τ. For further details the reader is referred to the journal article.

Aitcheson and Forrest (1994) extended DePaolo's equations and showed how they could be solved to find the fraction of material assimilated and values of P,

---

[1] In their Eq. (5), derived from Eq. (5.104) here, there is a misprint: the characters $a - 1$ should read $(a - 1)$.
[2] In their Eq. (1) the exponent of F should be $-z$ in both cases.

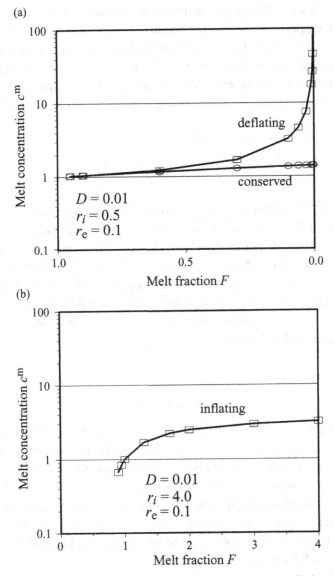

Fig. 5.12 Differing behaviour of an incompatible element in a deflating (a) and an inflating (b) magma chamber, calculated from Eq. (5.114) to illustrate the theory in Caroff *et al.* (1997). The trend for a magma chamber of fixed size is shown for comparison.

making extensive use of graphical methods. Roberts and Clemens (1995a,b) showed that, although AFC modelling of isotopic variations has been successful in some cases, the trace and major element concentrations often do not conform. Cavazzini (1996) showed how the degree of magma contamination may be determined at any point during an AFC process.

Cribb and Barton (1996) examined the consequences of assuming that assimilation is not proportional to crystallisation, i.e. with variable values of P, and generated a theory which they labelled as FCA, to 'denote decoupled fractional crystallisation and assimilation' (p. 294). Essentially, the element concentration is determined by a fractional crystallisation followed by mixing with the assimilant. Their Eq. (4) may be written as follows, using the symbols of the preceding sections

$$c^l L = c_f (L_0 - W) + c_a W_r \qquad (5.117)$$

where c_f is the concentration obtained using Rayleigh fractionation (Eq. 3.18). This relationship clearly describes a process of mixing of two independent sources. However, the authors do not describe this FCA process as *mixing*, and restrict this term to *magma mixing* which is used to describe 'the mixing of the evolved products of one magma batch with a new or later batch of primitive (or parental) magma, and therefore does not refer to mixing of magma with crustal melt (i.e. to assimilation)' (p. 299). This restriction confuses the subsequent discussion, because it does not conform to other authors' usage.

5.9 Summary

Resorption of previously crystallised minerals or assimilation of extraneous material by a magma may proceed by melting or by reaction, while cumulate minerals are crystallising. In either case energy is supplied from the magma. Subsequent homogenising is assumed to occur. Depending on the masses involved, *either* the magma will incorporate all available material, *or* will be used up leaving some material unassimilated. Trace element behaviour depends critically on assumptions about equilibration of the material being assimilated, but otherwise follows a modified Rayleigh fractionation.

Repeated eruption of similar lavas at a volcanic centre and repeated mineral layering in ultramafic coarse-grained sequences suggest that source magma chambers may undergo recharging from time to time with parent magma. Modelling the evolution of such a chamber, where fractionation, eruption and assimilation are combined with periodic replenishment, requires assumptions about the sequence of events, and these assumptions influence trace element behaviour; in some circumstances a steady state is reached, but in others a modified Raleigh fractionation takes place.

References

Aitcheson, S. J. and A. H. Forrest (1994) Quantification of crustal contamination in open magmatic systems. *Journal of Petrology*, **35**, 461–88.

Albarède, F. (1985) Regime and trace element evolution of open magma chambers. *Nature*, **318**, 356–8.

  (1995) *Introduction to Geochemical Modeling*. Cambridge: Cambridge University Press.

Caroff, M. (1997) Open system crystallization and mixing in two-layer magma chambers. *Lithos*, **36**, 85–102.

Caroff, M., Y. Lagabrielle, P. Spadea and J.-M. Auzende (1997) Geochemical modelling of non-steady state magma chambers: a case study from an ultrafast spreading ridge, East Pacific Rise, 17–19 °S. *Geochimica et cosmochimica acta*, **61**, 4367–74.

Cavazzini, G. (1996) Degrees of contamination in magmas evolving by assimilation-fractional crystallization. *Geochimica et cosmochimica acta*, **60**, 2049–52.

Cribb, J. W. and M. Barton (1996) Geochemical effects of decoupled fractional crystallization and crustal assimilation. *Lithos*, **37**, 293–307.

DePaolo, D. J. (1981) Trace element and isotopic effects of combined wallrock assimilation and fractional crystallization. *Earth and Planetary Science Letters*, **53**, 189–202.

Ehlers, E. G. (1972) *The Interpretation of Geological Phase Diagrams*. San Francisco: W. H. Freemam.

Hawkesworth, C. J., S. Blake, P. Evans *et al.* (2000) Time scales of crystal fractionation in magma chambers – integrating physical, isotopic and geochemical perspectives. *Journal of Petrology*, **41**, 991–1006.

Irvine, T. N. and C. H. Smith (1967) The ultramafic rocks of the Muskox intrusion, Northwest Territories, Canada. In *Ultramafic and Related Rocks*, ed. P. J. Wyllie. New York: John Wiley & Sons, pp. 38–49.

O'Hara, M. J. (1977) Geochemical evolution during fractional crystallisation of a periodically refilled magma chamber. *Nature*, **266**, 503–7.

  (1995) Trace element geochemical effects of integrated melt extraction and 'shaped' melting regimes. *Journal of Petrology*, **36**, 1111–32.

O'Hara, M. J. and N. Fry (1996) The highly compatible trace element paradox – fractional crystallisation revisited. *Journal of Petrology*, **37**, 859–90.

O'Hara, M. J. and E. Matthews (1981) Geochemical evolution in an advancing, periodically replenished, periodically tapped, continuously fractionated magma chamber. *Journal of the Geological Society of London*, **138**, 237–77.

Pankhurst, R. J. (1977) Open system crystal fractionation and incompatible element variation in basalts. *Nature*, **268**, 36–8.

Reagan, M. K., J. B. Gill, E. Malavassi and M. O. Garcia (1987) Changes in magma composition at Arenal volcano, Costa Rica, 1968–85: real-time monitoring of open-system differentiation. *Bulletin of Volcanology*, **49**, 415–34.

Roberts, M. P. and J. D. Clemens (1995a) The perils of AFC modelling (abstr). In *The Origin of Granites and Related Rocks*, ed. M. Brown and P. M. Piccoli. United States Geological Survey Circular 1129, p. 126.

  (1995b) Feasibility of AFC models for the petrogenesis of calc-alkaline magma series. *Contributions to Mineralogy and Petrology*, **121**, 139–47.

Usselman, T. M. and D. S. Hodge (1978) Thermal control of low pressure fractionation processes. *Journal of Volcanology and Geothermal Research*, **4**, 265–81.

6

Trace element evidence for crystallisation processes

6.1 Introduction

In the preceding chapters the behaviour of trace element concentrations has been modelled for various kinds of magma crystallisation. In almost all of these cases the prime variable has been the fraction of residual liquid, F, or the degree of crystallisation $(1 - F)$. The same approach will be used in subsequent chapters to model element trends during melting processes.

The current chapter will, however, look into ways in which crystallisation processes can be detected from element data.

6.2 Variation diagrams

When faced with the interpretation of a set of related igneous rock analyses, to see whether any of the crystallisation models is appropriate, the most immediate difficulty is that the values of F are never known.[1]

It may be possible to use petrographic features and major element data to rank the rocks in degree of evolution from most primitive to most evolved, which would correspond to a series of decreasing F-values. A long-established method of this kind is to use the silica content as a measure of the degree of evolution and then plot the elements of interest against it. This gives a *variation diagram*, often referred to by its inventor as a *Harker diagram*. Such diagrams were used extensively by Bowen (1928) to illustrate the *liquid line of descent* during igneous differentiation. Various other element combinations have been introduced in place of silica, one of which was the Larsen index

$$\frac{1}{3}SiO_2 + K_2O - (CaO + MgO) \tag{6.1}$$

---

[1] An approximate method of measuring F-values for a series of samples, using a strongly incompatible element, will be discussed shortly.

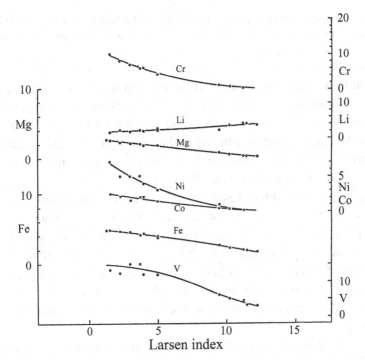

Fig. 6.1 Variation diagrams showing elemental abundance (wt. % or ppm) of Mg, Fe, Cr, Li, Ni, Co and V, plotted against Larsen index, for a series of rock samples from Crater Lake, Oregon (Nockolds and Allen, 1953, Fig. 2).

Figure 6.1 shows several elements plotted against the Larsen index for a series of rocks from Crater Lake, Oregon (Nockolds and Allen, 1953). Smooth trends were taken to indicate samples of a series of liquids formed by differentiation from a common parent, lying somewhere beyond the left-most point. Figure 6.1 shows smooth curves for the major elements and for the trace elements Cr, Li and Co, but there is some scatter for Ni and V.

Another measure used is the *differentiation index* of Thornton and Tuttle (1960) which consists of the sum of the hypothetical quartz + feldspar + feldspathoid minerals calculated from a rock analysis, since these minerals represent the residual materials to which igneous differentiation leads (*petrogeny's residual system*) and which therefore should accumulate in more evolved liquids.

Such diagrams, however, are subject to a number of drawbacks:

(i) analytical errors may be large enough to conceal any trend;
(ii) an insufficient number of samples will give ambiguous results;
(iii) the rock samples may not represent liquids, i.e they may contain inclusions, or cumulate minerals, or may have suffered alteration;

(iv) there is a major theoretical problem, which is that concentrations are expressed in % oxides or elements; that is, they are numbers which have a constant sum or, in statistical terms, form a *closed array.*

In practical terms, this last means that the concentrations are not independent – e.g. if a rock contains 75% SiO_2 then no other component can exceed 25%. A diagram which uses Si or SiO_2 as the abscissa, with Ca or some other abundant element as the ordinate is almost bound to show a smooth negative slope. This was first studied in detail by Chayes (1960, 1971) but has subsequently been given much attention (see for example, Aitchison, 1986).

One way to circumvent the closed array drawback is to work with concentration ratios. Pearce (1968, 1970) shows that using the ratios of each oxide to one which is known to be, or presumed to be, constant can be much more useful; this is a Pearce element ratio (PER) diagram. The act of dividing concentration A by the concentration of component C, which has remained constant, in effect changes A to an *extensive variable*, not subject to the closed array drawback. 'The unique properties of PER analysis allow us to recover the "lost" information' (Pearce, 1990, p. 99.)

For instance, in Fig. 6.2, the left-hand portion (a) shows a conventional Harker diagram for analyses of the Palisade sill, New Jersey, which crystallised from a basaltic parent melt (P) with a lower zone enriched in olivine; this diagram shows regular variation of the main oxides but does not suggest any mechanism which might have been operating. The right-hand portion (b), however, shows the effect of dividing each oxide by the Al_2O_3 concentration, which is presumed to have remained constant while olivine was crystallising; the diagram shows clearly that only FeO, MgO and SiO_2 were being extracted from the system, 'and further, that the rates at which they are extracted are determined by the stoichiometry of olivine' (Pearce, 1968, p. 150).

The advantage of this kind of variation diagram is clear, but it is important to realise that the choice of the constant component is constrained by the model being tested. The model is simple in Fig. 6.2, but if the crystallising assemblage comprises two or three solid-solution series it becomes difficult to find a quasi-constant component. Such problems are investigated in the later paper, which shows some volcanic series to be rather intractable (Pearce, 1990).

6.3 Other two-element plots

Variation diagrams or PER diagrams are most often used to display major-element behaviour, where closed-array constraints are important. Similar diagrams presenting minor and trace element data are widely used in studying igneous rock series.

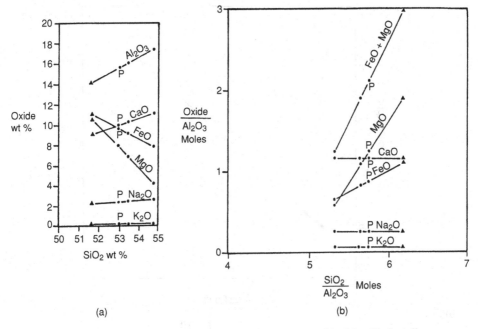

Fig. 6.2 Data from the Palisade sill, New Jersey, are used in (a) a Harker diagram and (b) a PER diagram (Pearce, 1968, Fig. 5); P is the presumed parent magma. The Harker diagram shows systematic changes in the % oxide values (*intensive variables*) but gives no information on the process taking place. Division by % Al$_2$O$_3$, which is presumed to remain constant, converts the oxides in (b) in effect to *extensive variables* and shows that the magma evolved by crystallising olivine, using Mg, Fe and Si, without any changes in the non-participating elements Ca, Na and K (as well as Al).

There are many good examples of correlation (linear or curved) between arithmetic plots of pairs of element analyses in igneous rocks of related parentage – examples where the analytical precision appears to have been good enough to not obscure relationships. Examples from Bowman *et al.* (1973) and Villemant *et al.* (1981) are shown in Fig. 6.3, and others are to be found in Gunn *et al.* (1970), Weaver *et al.* (1972), Helmke and Haskin (1973), Paster *et al.* (1974), Ferrara and Treuil (1975) and Treuil and Joron (1975). The Bowman *et al.* analyses of Californian obsidians and dacites (Fig. 6.3a) show an exceptional degree of linear correlation between Sc and Fe, and the linear trend does not pass through zero; similar behaviour is shown by several other elements, both incompatible and compatible. In other cases, such as the Villemant *et al.* study of the Chaine des Puys alkali basalts (Fig. 6.3b) good straight line correlations between U–Th and Rb–Th are seen, but in this case the lines pass through the origin; compatible elements such as Co and Sc show curved relationships with Th. Inflections or nick-points occur in some such

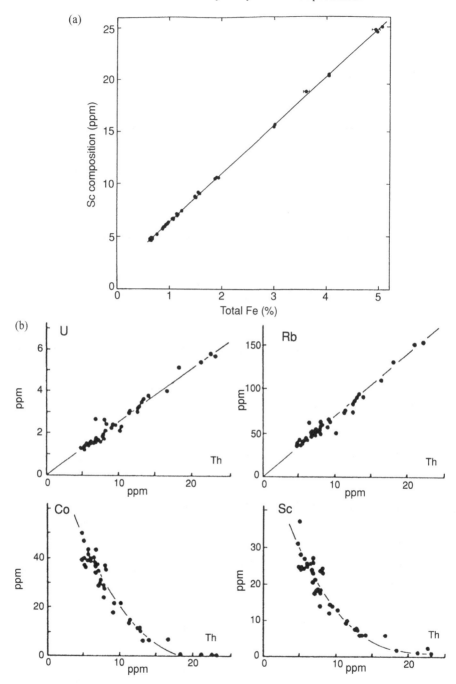

Fig. 6.3 (a) Sc concentration plotted against total Fe in 43 obsidian and dacite samples, Borax Lake, California (Bowman *et al.*, 1973, Fig. 2). There is a strong linear correlation but since it does not pass through the origin it is more likely that the lavas originated by some process of magma mixing, rather than crystallisation differentiation. (b) Element pairs from analyses of samples from an alkali basalt series from the Chaine des Puys, Massif Central, France (Villemant *et al.*, 1981, Fig. 1). The two incompatible elements, U and Rb, show strong linear correlation with Th, passing through the origin; by contrast, the compatible Co and Sc show good curved correlation.

diagrams, and indicate some change in the control process – usually a change in the minerals crystallising.

These examples come from fresh volcanic rock series and, except where porphyritic, can be approximated to samples of evolving liquids. The controls over such evolution include:

(i) crystallisation (equilibrium or fractional);
(ii) mixing;
(iii) assimilation;
(iv) melting.

Melting will be considered further on, but the first three have been considered in previous chapters. Interpretations of such element–element diagrams has been made by the authors already quoted and, for the related case of lunar rocks, by Haskin *et al.* (1970) and Helmke *et al.* (1972).

6.3.1 Crystallisation trends

The *equilibrium crystallisation* equation (Eq. 3.12) for an element whose partition coefficient is D_α, can be written

$$F = \frac{c_{\alpha 0} - c_\alpha^L D_\alpha}{c_\alpha^L (1 - D_\alpha)} \tag{6.2}$$

Combining this with a similar expression for the concentration c_β^L of a second element gives the following relationship

$$c_\beta^L = r \cdot c_\alpha^L \cdot \frac{c_{\beta 0}}{c_{\alpha 0}} + c_\alpha^L c_\beta^L \cdot \frac{1 - r}{c_{\alpha 0}} \tag{6.3}$$

where

$$r = \frac{1 - D_\alpha}{1 - D_\beta} \tag{6.4}$$

Equation (6.3) may be recast to give

$$\frac{1}{c_\alpha^L} = \frac{p}{c_\beta^L} + q \tag{6.5}$$

where p and q are constants. So, as a consequence, pairs of samples from such a relationship in a plot of $1/c_\beta^L$ vs. $1/c_\alpha^L$ will show a linear trend (see Fig. 6.4a) and c_β^L vs. c_α^L or log c_β^L vs. log c_α^L a curved trend.

(a)

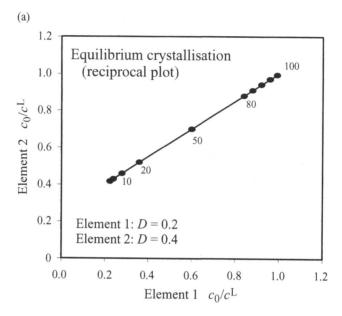

(b)

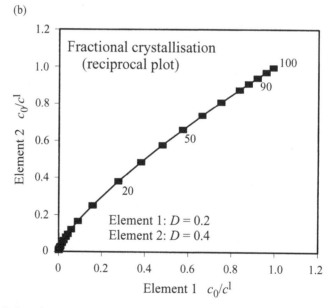

Fig. 6.4 Pairs of reciprocal concentrations of slightly incompatible trace elements in samples of liquids which have evolved by equilibrium (a) or fractional (b) crystallisation show straight and curved trends respectively. If the concentrations are relative, as is the case here, the trends pass through the point (1,1).

Table 6.1 *Characteristics of two-element plots for trace elements in magmatic rocks*

	Equilibrium crystallisation	Fractional crystallisation	Two-source mixing[a]
c_β vs. c_α	curved	curved	linear
$\log c_\beta$ vs. $\log c_\alpha$	curved	linear	curved
$1/c_\beta$ vs. $1/c_\alpha$	linear	curved	curved
c_β/c_α vs. c_β	linear	curved	curved
$\log (c_\beta/c_\alpha)$ vs. $\log c_\beta$	curved	linear	curved

[a] Plot does not pass through origin.

The *Rayleigh fractional crystallisation* equation (Eq. 3.20) for element α can be written

$$F = \left(\frac{c_\alpha^1}{c_{\alpha0}} \right)^{\frac{1}{(D_\alpha - 1)}} \tag{6.6}$$

and, when combined with a similar equation for element β, gives the following

$$\frac{c_\alpha^1}{c_{\alpha0}} = \left(\frac{c_\beta^L}{c_{\beta0}} \right)^r \tag{6.7}$$

or

$$\log c_\alpha^1 = r \cdot \log c_\beta^1 + \log c_{\alpha0} - r \cdot \log c_{\beta0} \tag{6.8}$$

As a consequence, in this case, pairs of samples plotted as $1/c_\beta^L$ vs. $1/c_\alpha^L$ or c_β^L vs. c_α^L will show a curved trend (Fig. 6.4b); and $\log c_\beta^L$ vs. $\log c_\alpha^L$ a linear trend.

These results are summarised in Table 6.1. It should be noted that each of the five plots is of relative concentrations and will pass through the origin, which is the unit point (1,1) whose coordinates correspond to $c_{\alpha0}$ and $c_{\beta0}$; of course in real rock series, $c_{\alpha0}$ and $c_{\beta0}$ are not known, so the trend lines will not pass through zero.

Equations (6.3) and (6.7) thus appear to give a method for deciding whether a set of analyses represent products of equilibrium or fractional crystallisation; two-element reciprocal plots will differ as shown in Figs. 6.4a and 6.4b, although the difference may not be apparent if there is a restricted range of values. However, this is only the case when the D-values are not close to zero. Figure 6.5 shows pairs of melt concentrations for fractional crystallisation where one element has a coefficient of 0.001, and the other is either 0.01 or 0.4; in the first case the plot is virtually linear, but this is not the case when the other element is only weakly incompatible.

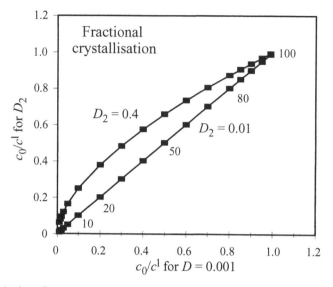

Fig. 6.5 A plot of pairs of incompatible trace element concentrations in samples of liquids which have evolved by fractional crystallisation shows near-linear trends, passing through the origin, when both of the elements are strongly incompatible (i.e. have $D \ll 1$). If one of the elements, however, is only mildly incompatible the plot will not be linear.

As pointed out first by Anderson and Greenland (1969), when $D \to 0$ then for both equilibrium and fractional crystallisation

$$F \approx \frac{c_{\alpha 0}}{c_{\alpha}^{l}} \qquad (6.9)$$

and is the reason for the near-linear plot in Fig. 6.5. The degree of discord between the true and the estimated melt fraction, expressed in terms of the percentage Δ, at different values of F (as given by Allègre *et al.*, 1977) is shown in Fig. 6.6, for three values of D

$$\Delta = 100 \cdot \left(\frac{\frac{1}{F} - F^{D-1}}{\frac{1}{F}} \right) \qquad (6.10)$$

It is evident from Fig. 6.6 that the approximation is only acceptable for values of D equal to or less than 0.01. Elements behaving in this way are described as 'hygromagmatophile' by Allègre *et al.* (1977), although this term was intended only to be synonymous with *incompatible* when first used (Treuil and Varet, 1973; Treuil and Joron, 1975).[2] It should not be forgotten, that the discussion centering

---

[2] Referring to their expression of Eq. (6.8), Treuil and Joron (1975, p. 132) observe that 'La relation . . . est d'autant mieux satisfaite que l'élément est plus hygromagmatophile, c'est-à-dire que D est petit'.

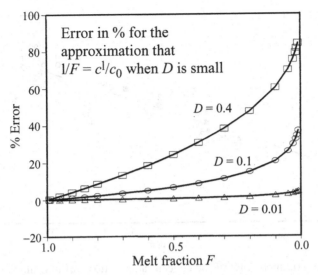

Fig. 6.6 The assumption that the relative concentration of a strongly incompatible element is equal to the reciprocal of F is approximate. The error incurred is shown here as a percentage of the actual value of F for three values of D (after Allègre et al., 1977, Fig. 1).

on Figs. 6.4 and 6.5 is only relevant for analysis sets of high precision, otherwise a spread of points in the plots may mask any choice between linear and curved trends.

6.3.2 Element ratio plots

In studies of trace elements in volcanic rock series, plots of a two-element ratio against its numerator, e.g. c_β^L/c_α^L vs. c_β^L, both arithmetic and logarithmic concentrations have been used (see Table 6.1). In the case of equilibrium crystallisation, Eq. (6.3) may be modified to give

$$\frac{c_\beta^L}{c_\alpha^L} = (1 - r) \cdot c_\beta^L + r \cdot \frac{c_{\beta 0}}{c_{\alpha 0}} \tag{6.11}$$

This will give a linear plot of the ratio against its numerator concentration. For fractional crystallisation, the relationship is readily derived from Eq. (6.8) and is as follows

$$\log \left(\frac{c_\beta^l}{c_\alpha^l} \right) = (1 - r) \cdot \log c_\beta^l + r \cdot \log c_{\beta 0} - \log c_{\alpha 0} \tag{6.12}$$

So a logarithmic plot will be linear, and will pass through the unit point (1,1), but an arithmetic plot will consequently be curved (hyperbolic).

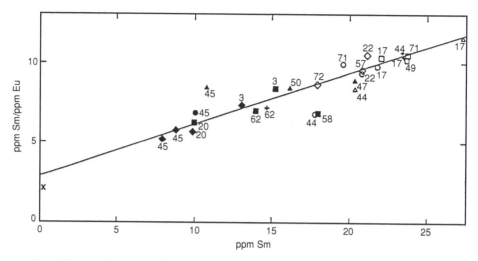

Fig. 6.7 An arithmetic plot of the concentration ratio Sm/Eu against Sm for numerous Apollo 11 basalts shows a linear trend (Haskin *et al.*, 1970, Fig. 6); this is also shown by a logarithmic plot, which led to an interpretation of origin by fractional crystallisation. The line is a least squares fit of all the points but three outliers.

Equation (6.12) was found by Haskin *et al.* (1970) to apply to the concentration ratio log [Sm/Eu] plotted against log [Sm] concentrations in numerous Apollo 11 basalts, leading to an interpretation that they formed by fractional crystallisation. The authors, however, found that an arithmetic plot was linear also (as in Fig. 6.7) which would be better interpreted as indicating fractional melting (see later).

6.3.3 Mixing and assimilation

The principal effects of magma mixing or magma assimilation on element concentration diagrams have been treated by Vollmer (1976), Langmuir *et al.* (1978), von Engelhardt (1989); the first two works are mainly concerned with isotopic ratios, and the last with elemental abundances.

If there are only two sources, or end-members, to be considered and the concentrations of four elements (or oxides) are a, b, c and d then the following kinds of plots are the most relevant:

element a vs. element b linear with intercepts
ratio a/b vs. ratio c/d hyperbolic
ratio a/b vs. element c hyperbolic
ratio a/b vs. element a hyperbolic

The mixing equations may be derived as follows. Let the end-members have concentrations a_1, b_1, c_1, d_1 and a_2, b_2, c_2, d_2. If any mixture has concentrations a, b, c and d, and is composed of fractions x_1 and x_2 of the end-members, then

$$a_1 x_1 + a_2 x_2 = a \tag{6.13}$$

$$b_1 x_1 + b_2 x_2 = b \tag{6.14}$$

$$c_1 x_1 + c_2 x_2 = c \tag{6.15}$$

$$d_1 x_1 + d_2 x_2 = d \tag{6.16}$$

and similar equations govern all other components.
   Also

$$x_1 + x_2 = 1 \tag{6.17}$$

$$x_1 = \frac{a - a_2}{a_1 - a_2} \tag{6.18}$$

So, if we construct the ratios $\alpha = a/b$ and $\beta = c/d$, then substitute from the above equations, the following relationship is obtained between α and β:

$$\beta(a_2 d_1 - a_1 d_2) + \alpha(b_2 c_1 - b_1 c_2) \\ + \alpha\beta(d_2 b_1 - d_1 b_2) + (c_2 a_1 - c_1 a_2) = 0 \tag{6.19}$$

which is a hyperbola. This may be modified to obtain the hyperbolas for the other kinds of plot by substituting for c or d.

   For example, Fig. 6.8 shows the linear relationships expected in plots between any two concentrations, where mixing of two sources has taken place, and the hyperbolic relationship where one of the variables is the ratio of two concentrations.

   An example of a mixing model is shown in Fig. 6.9a, using the elements Cs and Co, and assuming that the two end-member magmas could be represented by the extreme compositions (i.e. the richest and the poorest) of the Borax Lake volcanics previously mentioned (Section 6.3.1), calculating the hyperbola from Eq. (6.19). Analyses of the additional volcanic rocks from the Borax Lake series (Bowman *et al.*, 1973) are plotted in Fig. 6.9b and show marked agreement with the model. The authors, nevertheless, found it difficult to accept that mixing could have taken place.

   If the mixing involves three or more end-members then ratios such as a/b and c/d will not give regular or predictable relationships.

6.4 Inversion modelling

Forward modelling is the customary approach to geochemical problems; in brief, this is the search for a hypothesis to explain the chemical composition of a suite of collected igneous rocks, making various assumptions, such as the composition of

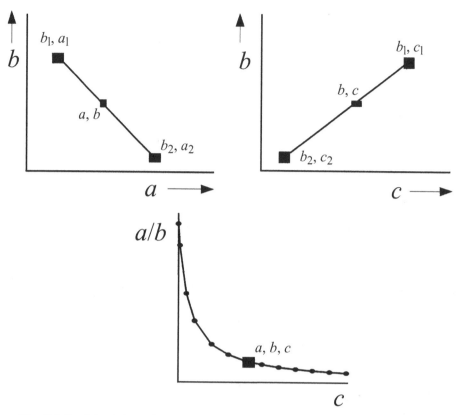

Fig. 6.8 Partial mixing of two magmas, containing % or ppm concentrations of elements a_1, b_1, c_1 . . . and a_2, b_2, c_2 . . . , will give a linear array of points on plots between pairs of concentrations, whether the elements are compatible or incompatible; the array does not pass through the origin, so the intercepts are non-zero. A plot of the ratio a/b of two concentrations against either of the concentrations or another (c) gives a hyperbola.

the source materials, the partition coefficients of the elements, the mechanisms of magma evolution etc. Many of the components of such modelling have been outlined in this and the preceding chapters.

In recent years attention has been given to the opposite approach, that of *inversion* of the data. In the case of trace elements the aim is to reconstruct the trace element path (TEP) in multidimensional space, i.e. to determine such things as the composition of the source materials, the partition coefficients, the degree of melting of the source, the extent of fractionation of a magma etc. The method depends on having more samples than variables (i.e. the problem is over-determined) and consequently a probabilistic solution can be found.

The introduction of inversion theory to geochemistry began with the publication of three papers by C. J. Allègre and co-workers (Allègre *et al.*, 1977; Minster *et al.*, 1977; Minster and Allègre, 1978); the first of these explained the principles and the

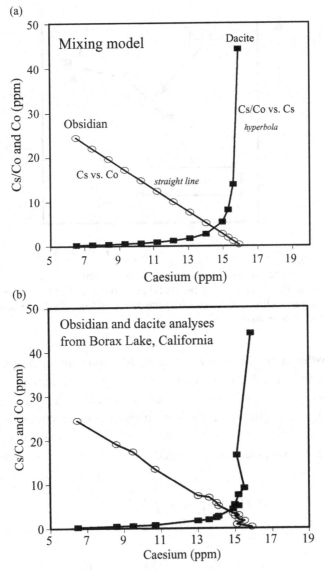

Fig. 6.9 (a) Mixing of magmas is modelled using the extreme compositions of obsidian and dacite from the Borax Lake volcanics (Bowman *et al.*, 1973, Tables 5 and 6); the former is high in Cs and low in Co, and the latter the opposite. (b) Analyses of the intermediate Borax Lake volcanics show a good fit to the mixing model in (a).

second and third explored the algebraic treatment and mathematical solutions. Other aspects of inversion analysis are to be found in Albarède (1983), Powell (1984), Zou (1997) and Barca *et al.* (1998).

As an example, suppose that a magma with initial concentrations c_i^0 for n trace elements, is crystallising to a group of minerals, in constant proportions, such that

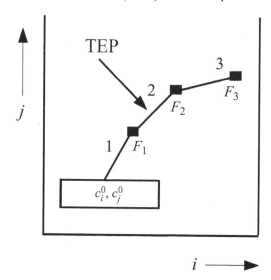

Fig. 6.10 Concentrations of two incompatible trace elements i and j, in a magma fractionating in three stages from an initial composition. The trace element path (TEP) is the composition variation with time, and is defined by the various parameters of the crystallisation.

it has bulk partition coefficients D_i. If melt is erupted in small quantities after successive fractions F_1, F_2, F_3 remain, then the melt concentration for element i takes on values c_i^1, c_i^2, c_i^3 and similarly for other elements. A plot for elements i and j might resemble Fig. 6.10, where the lines joining the initial composition to points 1, 2, 3 form a projection of the multidimensional TEP. If it is assumed that Raleigh fractionation is taking place, then the successive concentrations can be expressed as follows

$$c_i^1 = c_i^0 f_1^{D_i - 1} \quad c_i^2 = c_i^1 f_2^{D_i - 1} \quad c_i^3 = c_i^2 f_3^{D_i - 1} \tag{6.20}$$

where

$$f_1 = \frac{F_1}{F_0} \quad f_2 = \frac{F_2}{F_1} \quad f_3 = \frac{F_3}{F_2} \tag{6.21}$$

and $F_0 = 1$, the initial value.

If we consider only the elements shown on Fig. 6.10, then we have

$$n = 2 \text{ elements}, \ \lambda = 3 \text{ samples and } n\lambda = 6 \text{ analyses}$$

and the TEP is defined by

$$n = 2 \text{ values of } c_i^0, \ \lambda = 3 \text{ values of } F, n = 2 \text{ values of } D_i, \text{ i.e. 7 parameters.}$$

Consequently there are 6 knowns and 7 unknowns, or

$$n\lambda < 2n + \lambda \tag{6.22}$$

and the TEP can not be determined unless some additional assumptions are made.

If now we consider a more general case, where the number of elements determined on the three samples is larger, then it becomes possible that the reverse is true, i.e.

$$n\lambda > 2n + \lambda \tag{6.23}$$

or

$$(n-1)\lambda > 2n \tag{6.24}$$

In this case, it is said that the problem is *over-determined* and it is possible to seek the TEP by some form of inversion or regression analysis, giving a probabilistic solution for the values of c_i^0, λ and n.

Many of the applications of inversion analysis have been to systems where partial melting of the Earth's mantle has left residual solids, and the melts have subsequently undergone fractional crystallisation. The parameters here include the mineral proportions melting and in the residue, and the degree of melting, in addition to those already discussed. This topic will be considered in a later chapter.

The mathematical procedures of inversion analysis will not be included here, but may be found in Albarède and Provost (1977) and in the papers already referred to.

6.5 Summary

Relationships between pairs of elements or oxides in igneous rock series can help interpret crystallisation processes. Variation diagrams plotted between major element concentrations may only show consequences of closed array data but, if the data are divided by a component which has remained constant during crystallisation, the consequent PER plot may shed light on the crystallisation history.

Plots of pairs of trace element concentrations are less subject to closed array constraints. Arithmetic, logarithmic and reciprocal plots may indicate processes of equilibrium crystallisation, fractional crystallisation, magma mixing or melting (the last to be treated later).

Relative concentrations of the most incompatible elements are equal to $1/F$ and plots of pairs of such elements are nearly linear and pass through the origin for any crystallisation process, unless magma mixing or assimilation has taken place.

Inversion modelling of trace element data for a suite of related volcanic rocks tries to reconstruct the TEP in multidimensional space i.e. to determine such things as the composition of the source materials, the partition coefficients, the degree of melting of the source, the extent of fractionation of a magma etc. The method depends on having more samples than variables (i.e. the problem is over-determined) and consequently a probabilistic solution can be found.

References

Aitchison, J. (1986) *The Statistical Analysis of Compositional Data*. London: Chapman and Hall.

Albarède, F. (1983) Inversion of batch melting equations and the trace element pattern of the mantle. *Journal of Geophysical Research*, **88**, 10573–84.

Albarède, F. and A. Provost (1977) Petrological and geochemical mass-balance equations: an algorithm for least-square fitting and general error analysis. *Computers and Geosciences*, **3**, 309–26.

Allègre, C. J., M. Treuil, J. F. Minster, B. Minster and F. Albarède (1977) Systematic use of trace element in igneous process. Part I: Fractional crystallization processes in volcanic suites. *Contributions to Mineralogy and Petrology*, **60**, 57–75.

Anderson, A. T. and L. J. Greenland (1969) Phosphorus fractionation diagram as a quantitative indicator of fractional crystallisation differentiation of basaltic liquids. *Geochimica et cosmochimica acta*, **33**, 493–506.

Barca, D., G. M. Crisci and G. A. Rabuerul (1988) Further developments of the Rayleigh equation for fractional crystallisation. *Earth and Planetary Science Letters*, **89**, 170–2.

Bowen, N. L. (1928) *The Evolution of Igneous Rocks*. Princeton: Princeton University Press.

Bowman, H. R., F. Asaro and I. Perlman (1973) On the uniformity of composition in obsidians and evidence for magmatic mixing. *Journal of Geology*, **81**, 312–27.

Chayes, F. (1960) On correlation between variables of constant sum. *Journal of Geophysical Research*, **65**, 4185–93.

　(1971) *Ratio Correlation: A Manual for Students of Petrology and Geochemistry*. Chicago: University of Chicago Press.

Engelhardt, W. von (1989) Remarks on 'ratio plots' and 'mixing lines'. *Geochimica et cosmochimica acta*, **53**, 2443–4.

Ferrara, G. and M. Treuil (1975) Petrological implications of trace element and Sr isotope distributions in basalt–pantellerite series. *Bulletin volcanologique*, **38**, 548–74.

Gunn, B. M., R. Coy-Yll, N. D. Watkins, C. E. Abranson and J. Nougier (1970) Geochemistry of an oceanite–ankaramite–basalt suite from East Island, Crozet Archipelago. *Contributions to Mineralogy and Petrology*, **28**, 319–39.

Haskin, L. A., R. O. Allen, P. A. Helmke *et al.* (1970) Rare earths and other trace elements in Apollo 11 lunar samples. *Proceedings 1st Lunar Science Conference, Geochimica et cosmochimica acta supplement*, **1**, 1213–31.

Helmke, P. A. and L. A. Haskin (1973) Rare-earth elements, Co, Sc and Hf in the Steens Mountain basalts. *Geochimica et cosmochimica acta*, **37**, 1513–29.

Helmke, P. A., L. A. Haskin, R. L. Korotev and K. E. Ziege (1972) Rare earths and other trace elements in Apollo 14 samples. *Proceedings 3rd Lunar Science Conference, Geochimica et cosmochimica acta supplement*, **3**, 1275–92.

Langmuir, C. H., R. D. Vocke, G. N. Hanson and S. R. Hart (1978) A general mixing equation with applications to Icelandic basalts. *Earth and Planetary Science Letters*, **37**, 380–92.

Minster, J. F. and C. J. Allègre (1978) Systematic use of trace elements in igneous processes, III. Inverse problem of batch partial melting in volcanic studies. *Contributions to Mineralogy and Petrology*, **68**, 37–53.

Minster, J. F., J. B. Minster, M. Treuil and C. J. Allègre (1977) Systematic use of trace elements in igneous processes, II. Inverse problem of fractional crystallization process in volcanic suites. *Contributions to Mineralogy and Petrology*, **61**, 49–78.

Nockolds, S. R. and R. Allen (1953) The geochemistry of some igneous rock series. *Geochimica et cosmochimica acta*, **4**, 105–42.

Paster, T. P., D. S. Schauwecker and L. A. Haskin (1974) The behaviour of some trace elements during solidification of the Skaergaard layered series. *Geochimica et cosmochimica acta*, **38**, 1549–77.

Pearce, T. H. (1968) A contribution to the theory of variation diagrams. *Contributions to Mineralogy and Petrology*, **19**, 142–57.

  (1970) Chemical variations in the Palisades sill. *Journal of Petrology*, **11**, 15–32.

  (1990) Getting the most from your data: applications of Pearce element ratio analysis. In *Theory and Application of Pearce Element Ratios to Geochemical Data Analysis*, ed. J. K. Russell and C. R. Stanley. Geological Association of Canada: Short Course Vol. 8, pp. 99–130.

Powell, F. (1984) Inversion of the assimilation and fractional crystallization (AFC) equations; characterization of contaminants from isotope and trace element relationships in volcanic suites. *Journal of the Geological Society of London*, **141**, 447–52.

Thornton, C. P. and O. F. Tuttle (1960) Chemistry of igneous rocks I. Differentiation index. *American Journal of Science*, **258**, 664–84.

Treuil, M. and J.-L. Joron (1975) Utilisation des éléments hygromagmatophiles pour la simplification de la modélisation quantitative des processus magmatiques. Exemples de l'Afar et de la dorsale médioatlantique. *Società Italiana di mineralogia e petrologia,* Rendiconti, **XXXI**, 125–74.

Treuil, M. and J. Varet (1973) Critères volcanologiques, pétrologiques et géochimiques de la genèse et de la différenciation des magmas basaltiques: exemple de l'Afar. *Bullétin de la société géologique de France 7ème série*, **15**, 401–644.

Villemant, B., H. Jaffrezic, J. L. Joron and M. Treuil (1981) Distribution coefficients of major and trace elements: fractional crystallization in the alkali basalt series of Chaine des Puys (Massif Central, France). *Geochimica et cosmochimica acta*, **45**, 1997–2016.

Vollmer, R. (1976) Rb–Sr and U–Th–Pb systematics of alkaline rocks: the alkaline rocks from Italy. *Geochimica et cosmochimica acta*, **40**, 283–95.

Weaver, S. D., J. S. C. Sceal and I. L. Gibson (1972) Trace element data relevant to the origin of trachytic and pantelleritic lavas in the East African rift system. *Contributions to Mineralogy and Petrology*, **36**, 181–94.

Zou, H. (1997) Inversion of partial melting through residual peridotites or clinopyroxenes. *Geochimica et cosmochimica acta*, **61**, 4571–82.

7

Melting: basic trace element modelling

7.1 Introduction

Although there is some truth in the saying that melting is the opposite of freezing, this is not a very helpful concept in geochemistry, for a number of reasons. In simple magmatic crystallisation, for example, the initial system is a single phase which is a melt; as cooling proceeds a point will be reached where the solidus is intercepted, a solid phase begins to crystallise and the system becomes heterogeneous. The behaviour of a trace element will be governed by one partition coefficient, between the solid and the melt.

But in rock melting the initial system is already a heterogeneous phase system, a multi-phase mixture of minerals, and as heating proceeds a liquidus will be encountered and a melt phase develops. A bulk partition coefficient, including several of the individual mineral–melt coefficients, is needed to plot the behaviour of a trace element, because those minerals participate in the melting reaction.[1] As with crystallisation, the melting process may take place with *equilibration* between melt and solids at all times, or *fractionation* may occur, where successive liquids form from the residual mineral assemblage without interaction with previous liquids. Constraints on the process from kinetic and other factors will be considered in due course.

7.2 Melting a heterogeneous rock

The beginning of melting is not determined by the bulk composition of the rock, so much as by the mineral assemblage present. When two or more minerals become unstable in contact with each other a reaction begins in which one of the products is liquid; this occurs at a eutectic point on the liquidus surface in the appropriate

---

[1] There is a sense in which *all* the minerals in the rock participate in every melting reaction, but this will be treated later.

Inner part Outer part Surface ↓

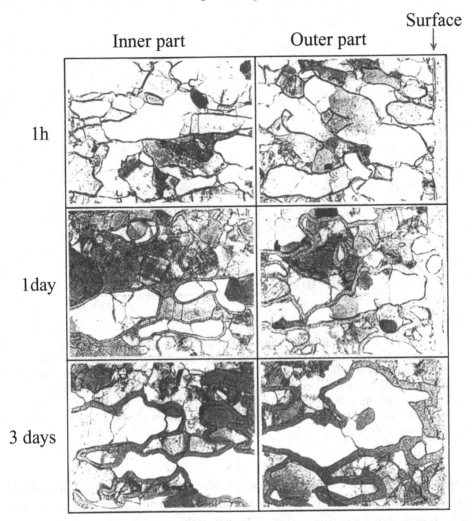

1h

1day

3 days

Fig. 7.1 Hydrothermal melting of a Norwegian gneiss for three periods at constant T and P (760°C, 2 kbar). The melt appears as dark bands around grain boundaries and its abundance increases with the melting time (Mehnert *et al.*, 1973, Fig. 2).

phase diagram. The reaction seldom consumes the minerals in the proportions in which they occur in the rock, i.e. it is said to be 'non-modal'; as a consequence, the mineral proportions p_j change during melting.

The mechanism of melting starts at grain boundaries, as shown in the experiments of melting a Norwegian gneiss under constant P and T for varying periods of time (Mehnert *et al.*, 1973). Layers of melt rock surround the grains participating in the reaction and become thicker as time goes on (Fig. 7.1). It may be inferred that after

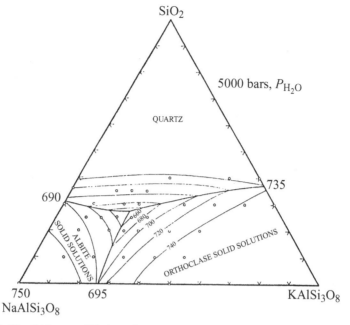

Fig. 7.2 The 5 kbar equilibrium diagram for the quartz–albite–orthoclase system, showing the compositions of H_2O-saturated liquids projected to the anhydrous base of the tetrahedron (after Luth *et al.*, 1964). Melting of any mixture of the three minerals will begin at the eutectic at $< 660\,°C$ at places where they have mutual contacts. Melting at two-mineral contacts, such as quartz–albite, albite–orthoclase or orthoclase–quartz would require higher temperatures of $690\,°C$, $695\,°C$ and $735\,°C$, respectively.

longer melting the rock would soon lose its coherence and collapse into a slurry. If such a rock be represented by a simple assemblage of quartz, albite and orthoclase, in the presence of water vapour (Fig. 7.2), then melting will begin at a grain contact of all three minerals. This will occur at a temperature of $\approx 650\,°C$, lower than any one of the contacts quartz–albite, albite–orthoclase, orthoclase–quartz, and the temperature will remain constant, as melting proceeds, until one of the minerals is exhausted. Clearly, this is an over-simplified account because the real system consists of solid solutions which influence the melting.

In a simpler system, the dry melting of a mixture q of two solid solutions in the simple binary system $X-Y$ (Fig. 7.3) begins at the eutectic temperature T_e. The two solid phases are x and y and they melt in the proportions ey/xe, which are different from their proportions ry/xr in the mixture q. As equilibrium melting proceeds the residual mineral proportion r approaches y, at which point x is all consumed and, while y continues to melt alone, the liquid composition moves from e up the liquidus towards T_{end}, where melting is complete. The bulk partition

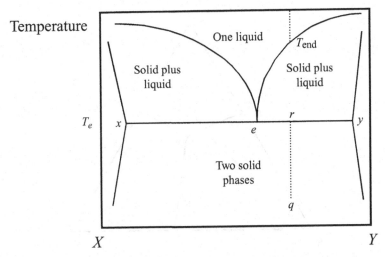

Fig. 7.3 Melting of a two-solid mixture q, in the hypothetical system X–Y will begin at the eutectic temperature T_e. The two solid solutions x and y melt, initially in the proportions ry/xy and xr/xy, at constant temperature. When phase x is exhausted the temperature rises, and more of y melts, until it is all consumed at T_{end}.

coefficient at the eutectic for some trace element i, present in both minerals, can be written

$$D_i = \frac{ry}{xy} D_i^{x-m} + \frac{xr}{xy} D_i^{y-m} \tag{7.1}$$

where the Ds are partition coefficients between each solid phase x or y and the melt. As the melting proceeds D_i will change, because the mineral proportions change.

   The details of the *melting path* in more complex systems depend on the system under consideration and whether equilibration or fractionation is in control, and have been examined in detail by D. C. Presnall in several publications (e.g. Presnall, 1969; Presnall and Bateman, 1973; Presnall, 1986). For equilibrium melting in a simple ternary system with no solid solution (Fig. 7.4), a mixture x begins to melt at the eutectic e. The temperature remains constant as the composition of the residual solid moves towards y. At this point all of phase C has been consumed and the residue is made up of A and B in the ratio By/yA. The temperature rises and melting continues as the components A and B melt in proportions given by the intersection with AB of the tangent to the cotectic curve. The liquid changes composition along ep and the residual solid along yB. When the liquid reaches p the system consists of liquid p and solid B, and the liquid changes along px while B melts. So the *liquid path* is epx and the *solid path* is xyB. Equilibrium crystallisation of a melt x follows exactly the opposite behaviour.

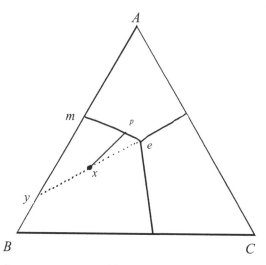

Fig. 7.4 A simple ternary system with no solid solution. Melting of a mixture of all three phases A, B and C, begins at the eutectic e and the detailed melting path is described in the text. This figure resembles Fig. 7.2 but differs in the presence of solid solutions.

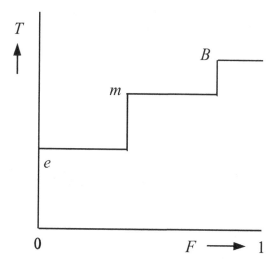

Fig. 7.5 Step-wise temperature changes during fractional melting in the system shown in Fig. 7.4, shown as a function of the melt fraction F.

Fractional fusion starts the same way with a eutectic liquid at e. But when phase C is exhausted, melting ceases because the residue y is isolated from the liquid and is at a temperature below the binary eutectic m. So heating continues until the residue recommences melting at m, and remains constant until A is consumed. Melting again ceases, until the temperature has risen to the melting point of B.

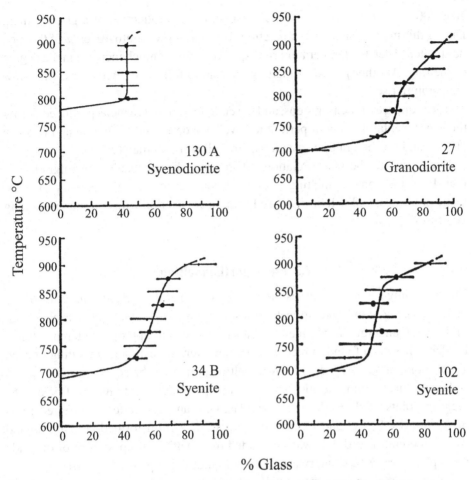

Fig. 7.6 Melting experiments on four rocks from the Debouillie stock, Maine, by Robertson and Wyllie (1971, Fig. 7) relate the degree of melting, measured by % glass produced, to the temperature, under a constant vapour pressure of 2 kbar. The bars show the uncertainty. The diagrams are in accord with Fig. 7.5, after allowing for the fact that natural minerals are solid solutions and thus melt over a temperature interval.

Thus the general principle in fractional fusion is that after one mineral is exhausted the temperature rises, without any melting, until the next invariant point is reached, and melting resumes (Fig. 7.5). This principle applies to more complex systems, except where the minerals are solid solutions, in which case the melting regions are not isothermal.

For example, in the experimental melting of silica-rich igneous rocks Robertson and Wyllie (1971) measured the approximate degree of melting, measured as % glass produced, as a function of T. The results (Fig. 7.6) show that melting begins at $< 700\,^\circ$C in three of the cases, corresponding to production of a granitic melt. During this melting step the temperature rises only slowly until 40–60% of the rock has melted. Then the temperature rises quickly without much increase in the degree of melting. Another period of melting appears to follow in three cases, with slow temperature rise.

The sequence of melting steps, and the composition of the melts produced during the experimental melting of peridotite have been described by Mysen and Kushiro (1976), and for the granitic system by Presnall and Bateman (1973).

In addition to the phase relations of melting, the mechanism of migration of melt from the zone of melting is of vital concern to the geochemistry of magmatic processes, and will be discussed later, in connection with continuous melting processes.

7.3 Early partial melting

The preceding section discusses the melting of mineral aggregates in terms of the thermodynamic stability relationships of solid phases of a somewhat ideal character, in that they undergo simple melting transformations at appropriate temperatures. Evidence has been found in recent years, however, indicating that melting of even simple compounds, with well-known melting points can be complex.

For instance, abnormal increases in enthalpy which are found well below the melting points of diopside, pseudowollastonite and anorthite are referred to as *premelting* by Richet and Fiquet (1991) and arise from formation of liquid and from structural disorder. It was concluded that melting of up to 20% of diopside takes place over a 44 K interval below the liquidus (Lange *et al.*, 1990).

Although this appears rather self-contradictory, *early partial melting* (EPM) of diopside and other minerals has also been reported by Doukhan (1995). For example, a diopside with about 2% of its Mg substituted by Fe, has a solidus at $1350\,^\circ$C, but heating to $1150\,^\circ$C generates very small droplets of silica-rich melt. These are initially too small for clear identification, but with continual annealing at $1200\,^\circ$C they coalesce into larger droplets which can be seen and analysed by X-ray microanalysis. It appears that the earlier view that diopside undergoes congruent melting is slightly incorrect, and it has in fact a field of incongruent melting, although only $13\,^\circ$C wide.

Doukhan (1995) also reports similar behaviour in enstatite and in olivine, under various ambient pressures and fO_2 levels, both in laboratory-prepared and natural phases. His assessment of the process is that 'EPM appears to be the early stage of conventional partial melting in the upper mantle. Partial melting would start by the nucleation of tiny droplets of EPM inside the individual grains of the various minerals . . . melt would be able to wet the grain boundaries and flow and escape through the network of grain boundaries' (Doukhan, 1995, p. 7).

It is possible that EPM may arise from traces of water, or other volatile components, adsorbed irregularly within the mineral. The liquidus of most minerals is lowered when water is present.

Even if EPM is commonly encountered in rock melting it is not clear that it would lead to trace element behaviour different from what is considered in the next sections. If, however, all minerals are susceptible to incongruous EPM, as seems to follow from the work by Doukhan and colleagues, this may influence the fractionation of incompatible elements.

7.4 Definitions

The following symbols, which are generally similar to or identical with those used in preceding chapters and defined in Chapter 3 will be used in this and succeeding chapters

W_0	initial mass of unmelted rock
L	mass of liquid or melt
$F = L/W_0$	fraction of liquid to total mass
W	mass of residual solid phases
W^f	mass of fluid or gas phase
W^i	mass of mineral i
X_i	mass fraction of mineral i in solid
p_i	mass fraction of mineral i in liquid
w_0	total mass of trace element
w^l or w^L	mass of trace element in liquid
w^s or w^S	total mass of trace element in residual solids
w^i	mass of trace element in mineral i
c_0	concentration of trace element in whole system
c^l or c^L	concentration of trace element in liquid
$\bar{c}$	concentration of trace element in accumulated liquid fractions
c^s	total concentration of trace element in solids
c_i	concentration of trace element in mineral i
D^{i-m}	partition coefficient for an element, between mineral i and melt or liquid m
D or D^{wr}	weighted or whole rock partition coefficient

In the following discussion it will initially be assumed that:

(i) partition coefficients D^{i-m} are constant;
(ii) melting proportions p_i are constant;
(iii) kinetic diffusion effects are ignored.

Conditions where these assumptions break down will be considered in due course.

7.5 Bulk partition coefficient

A rock undergoing melting consists of minerals 1, 2, 3, . . . , i, each in mass proportion X_i after a fraction F has melted. The trace element concentration in each is c_i and its partition coefficient with the melt is $D^{i-\mathrm{m}}$. Then, assuming that the minerals are in equilibrium with the melt, the concentration in the rock is

$$c^{\mathrm{S}} = \sum_i c_i X_i = c^{\mathrm{L}} \sum_i D^{i-\mathrm{m}} X_i \tag{7.2}$$

The total mass in the system is W_0, where

$$W_0 = L + W \quad L = F W_0 \quad W = (1 - F) W_0 \tag{7.3}$$

Also, if p_i is the proportion of mineral i in the melt, and X_i^0 the mineral proportion in the unmelted rock, then

$$W X_i = W_0 X_i^0 - L p_i \tag{7.4}$$

whence

$$X_i = \frac{X_i^0 - p_i F}{1 - F} \tag{7.5}$$

and so from Eqs. (7.2) and (7.5)

$$c^{\mathrm{S}} = c^{\mathrm{L}} \sum_i \frac{X_i^0 D^{i-\mathrm{m}} - p_i D^{i-\mathrm{m}} F}{1 - F} \tag{7.6}$$

If now we write $D_0 = \sum_i X_i^0 D^{i-\mathrm{m}}$ and $P = \sum_i p_i D^{i-\mathrm{m}}$ then we can define

$$D = \frac{D_0 - P F}{1 - F} \tag{7.7}$$

where D is the *bulk partition coefficient*.

A number of examples in the following pages will use values of $D^{i-\mathrm{m}}$ drawn from Table 7.1.

7.6 Trace elements in equilibrium melting

An alternative name for this process is *batch melting*. The mass balance for the trace element may be written

$$w_0 = w^{\mathrm{L}} + w^{\mathrm{S}} \tag{7.8}$$

or

$$c_0 W_0 = c^{\mathrm{L}} L + c^{\mathrm{S}} W$$

Table 7.1 *Selected partition coefficient values*

	ol	opx	cpx	sp	gar	amp	pl
La	0.000 053	0.000 4	0.053 6	0.000 02	0.001	0.12	0.034 8
Ce	0.000 105	0.001	0.085 8	0.000 03	0.004	0.18	0.027 8
Pr	0.000 251	0.002	0.137	0.000 1	0.02	0.3	0.022
Nd	0.000 398	0.004 1	0.187 3	0.000 2	0.057	0.45	0.017 9
Sm	0.000 7	0.006	0.291	0.000 4	0.14	0.6	0.013 2
Eu	0.000 8	0.01	0.329	0.000 6	0.26	0.7	0.3
Gd	0.001 5	0.012	0.367	0.000 9	0.498	0.7	0.012 5
Tb	0.002 1	0.016	0.405	0.001 2	0.75	0.66	0.011 6
Dy	0.002 7	0.02	0.442	0.001 5	1.06	0.63	0.011 2
Ho	0.005	0.026	0.415	0.002 3	1.53	0.6	0.011 4
Er	0.01	0.033	0.387	0.003	2.0	0.58	0.011 6
Tm	0.016	0.045	0.409	0.003 8	3.0	0.55	0.014
Yb	0.027	0.055	0.43	0.004 5	4.03	0.53	0.016
Lu	0.03	0.07	0.433	0.005 3	5.5	0.51	0.018
Y	0.006 5	0.009 6	0.467	0.004	3.08	0.46	0.011 5
Cr	1.8	2.8	15.0	600.0	13.0	6.0	0.04
Co	3.8	3.2	1.5	40.0	1.9	3.8	0.1
Ni	13.0	6.6	4.0	16.0	0.8	12.0	0.26
Ti	0.015	0.082	0.38	0.15	0.63	0.69	0.04
Zr	0.001	0.021	0.16	0.06	0.5	0.07	0.009 2
Hf	0.002 9	0.023	0.31	0.05	0.3	0.14	0.01
Sr	0.000 15	0.075	0.12	0.000 03	0.006 5	0.57	2.5
Rb	0.000 18	0.000 6	0.004 7	0.0	0.000 7	0.1	0.1

The coefficients were selected from listings by Allegre *et al.* (1977), Johnson *et al.* (1990), Gurenko & Chaussidon (1995), Ozawa and Shimizu (1995).

Mineral abbreviations refer respectively to olivine, orthopyroxene, clinopyroxene, spinel, garnet, amphibole and plagioclase.

So, first considering the (unrealistic) case of *modal melting*, where $X_i^0 = p_i$ and $D_0 = P = D$ then substituting $c^S = D\,c^L$ leads to

$$c^L = \frac{c_0}{D + F - DF} \tag{7.9}$$

$$c^S = \frac{D\,c_0}{D + F - DF} \tag{7.10}$$

Next, for *non-modal melting*, using Eq. (7.7), it is readily found that

$$c^L = \frac{c_0}{D_0 + F(1 - P)} \tag{7.11}$$

$$c^S = \frac{D_0 - PF}{(1 - F)(D_0 + F(1 - P))} \tag{7.12}$$

Table 7.2 *Calculation of* F_{max} *and* F_b *from Eqs.* (7.22) *and* (7.23) *using hypothetical mineral proportions for melting a garnet lherzolite*

	$D^{Yb-melt}$	Mode X_i^0	Melt p_i	F_{max}	F_b
olivine	0.027	0.3			
orthopyroxene	0.055	0.2			
clinopyroxene	0.43	0.3	0.5	0.6	
garnet	4.03	0.2	0.5	0.4	
D_0		0.95			
P			2.23		0.43

Partition coefficients from Table 7.1.

In using Eqs. (7.11) and (7.12) it should be realised that they have bounds in addition to the limiting value of $F = 1$. One of these comes from Eq. (7.5) and is the value of F for which the numerator equals zero. This condition must be tested for each mineral which is melting, and when the melt fraction reaches

$$F_{max} = \frac{X_i^0}{p_i} \tag{7.13}$$

mineral i is exhausted and the melting, if it continues, will have new mineral proportions and a new value of D.

Another bound comes from the bulk partition coefficient, and is the the value F_b for which the numerator of Eq. (7.7) becomes zero, i.e. when

$$F_b = \frac{D_0}{P} \tag{7.14}$$

Usually this restriction is not important, because some mineral will be exhausted first. But Eq. (7.14) imposes a limit to arbitrary modelling choices of P because, since F_b cannot exceed unity, so P must be chosen as $\leq D_0$.

An example of the limiting bounds for F is given in Table 7.2. With the mineral proportions shown, garnet will all have gone when 40% of the rock has melted; at this point the constraint or limit from the value of P at 43% has not yet been reached.

7.7 Trace elements in fractional melting

Following Rayleigh (1902), it is assumed that a rock is melting, so that in a small length of time

$$\text{mass of melted solid} = dW$$
$$\text{mass of trace element lost from solid} = dw^s = c^s dW$$

and so
$$\text{mass of residual solid} = W - dW$$
$$\text{mass of trace element left in residual solid} = w^s - dw^s$$

Consequently the concentration of the trace element in the instantaneous drop of melt is

$$c^l = \frac{dw^s}{dW} = \frac{d(c^s W)}{dW} \tag{7.15}$$

and, if we assume that c^l is some function of c^s, say $f[c^s]$, then

$$\frac{d(c^s W)}{dW} = f[c^s] \tag{7.16}$$

whence, since $dW = -dF \cdot W_0$

$$-\frac{dF}{1 - F} = \frac{dc^s}{f[c^s] - c^s} \tag{7.17}$$

In the case of *modal melting*, $f[c^s]$ is linear with c^s, and in order to integrate we substitute $f[c^s] = c^l = c^s/D$, which gives

$$-\frac{dF}{1 - F} = \frac{dc^s}{c^s \left(\frac{1}{D} - 1\right)} \tag{7.18}$$

which can be integrated to give

$$c^s = c_0 (1 - F)^{(\frac{1}{D} - 1)} \tag{7.19}$$
$$c^l = c^s/D \tag{7.20}$$

In this formulation, D is invariable, which can seldom apply to real rocks.

In non-modal melting the minimum melting mixture is different from the mode so $p_i \neq X_i$ for all i, and consequently D follows Eq. (7.7). The $f[c^s] - c^s$ relation is now no longer linear because

$$f[c^s] = c^s \cdot \frac{1 - F}{D_0 - PF} \tag{7.21}$$

but melting is still governed by Eq. (7.18), which may then be integrated to give

$$c^s = \frac{c_0}{1 - F} \cdot \left(1 - \frac{PF}{D_0}\right)^{\frac{1}{P}} \tag{7.22}$$

and

$$c^l = \frac{c_0}{D_0} \cdot \left(1 - \frac{PF}{D_0}\right)^{(\frac{1}{P} - 1)} \tag{7.23}$$

The concentration averaged over *all the melt* so far produced is readily found to be

$$\bar{c} = \frac{c_0}{F} \cdot \left(1 - \left(1 - \frac{PF}{D_0} \right)^{\frac{1}{P}} \right) \tag{7.24}$$

An alternative formulation of the melting equation starts with the definition of the concentration in the residual solid

$$c^s = \frac{w_0 - w^l}{W_0 - L} \tag{7.25}$$

Then the momentary concentration in the liquid is given by $c^l = dw^l/dL$, and if this is in equilibrium with the residual solid then

$$c^l = \frac{dw^l}{dL} = \frac{1}{D} \cdot \frac{w_0 - w^l}{W_0 - L} \tag{7.26}$$

and

$$\frac{dw^l}{w_0 - w^l} = \frac{1}{D} \cdot \frac{dF}{1 - F} \tag{7.27}$$

This was given as Eq. (7) in Shaw (1970) and is in some circumstances more useful than Eq. (7.18).

The concentrations obtained from equilibrium melting (Eqs. 7.11 and 7.12) and fractional melting (Eqs. 7.22 and 7.23) are plotted in Figs. 7.7a and 7.7b using large and small values of D_0, choosing appropriate values of P, to illustrate several points. The x-axis has been expanded for low values of F by using logarithmic values.

The equilibrium plot shows the melt concentrations converging towards unity for any value of D_0; when $F = 1$ the rock must be completely molten. Also, the concentration paths are seen to be very similar for small degrees of melting, and the trends only begin to diverge when the melting degree exceeds 10%. For an incompatible element ($D_0 = 0.10$) the first fraction of melt produced is markedly enriched, but after further melting the concentration drops off quickly. A compatible element ($D_0 = 4$), however, is retained in the unmelted solid residue over much of the melting range, little going into the melt until a substantial fraction of the rock is molten.

The fractionation plots are virtually identical to the equilibrium ones for up to 5% melting, and only diverge to a major degree when 20% or more of the rock is melted.

The importance of partial melting in controlling the production of basaltic magma from the mantle was the main theme of an important paper by Paul Gast (1968).

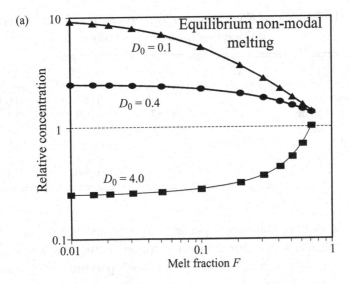

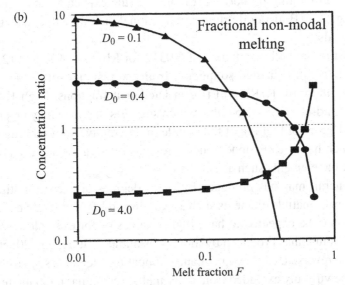

Fig. 7.7 (a) Trace element melt concentrations during equilibrium (batch) non-modal melting of multi-phase solids with bulk partition coefficients of 0.1, 0.4 and 4.0. The value of P in each case is equal to $D_0/0.75$, so that melting ends at $F_b = 0.75$ and would continue with a different mineral assemblage until the mass is completely liquid at $F = 1.0$. (b) Similar diagram to (a) but for fractional melting. In this case the melt relative concentrations show strong fractionation trends towards the end of melting.

Table 7.3 *Rb and K partition coefficients and peridotite melting proportions (from Gast, 1968)*

	X_i^0	p_i	D_{Rb}^{i-m}	D_K^{i-m}
olivine	0.55		0.0044	0.0067
orthopyroxene	0.2	0.167	0.044	0.1
amphibole	0.2	0.833	0.1	0.5
spinel	0.05		0.0	0.0
D_0			0.0312	0.1237
P			0.0906	0.4332

The behaviour of trace elements depends in the first instance on their abundance in the source material, and then on their partition coefficients. In the case of an incompatible element such as Rb, then using Gast's partition coefficients, mineral proportions and melting proportions (Table 7.3) the expected behaviour of Rb is shown in Fig. 7.8a. Both equilibrium and fractional melting will concentrate Rb in the early liquids.

In contrast to the bulk coefficient of 0.0312 for Rb, that of K is 0.1237, which means that K is not extracted so quickly from the residual solid (Fig. 7.8b) and consequently the ratio K/Rb will rise in later melt fractions. High K/Rb ratios are found in mid-ocean ridge basalts, which suggests that they may be products of substantial degrees of melting of the mantle or, alternatively, of melting of mantle material which has already undergone a previous melting event which depleted it of incompatible elements such as Rb.

So, in general, magmas produced by small degrees of partial melting will be enriched in incompatible elements such as Rb, Ba and LREE,[2] relative to K, Sr and HREE, and may be expected to have higher values of for example Rb/K, Ba/Sr and La/Yb. In magmas produced by more substantial melting, e.g. 20–50%, these trends will be reversed. On the other hand, compatible elements such as Ni, Co, Sc and Y, will be virtually excluded from early melts, accumulating in the residue and in later melts.

These characteristics suggested to Gast that this process could account for the origins of many basaltic series, especially the oceanic ones. The early melt fractions could be identified as the ocean island alkalic basalts, with the ocean ridge tholeiitic basalts coming from more extended melting, as already mentioned. More recent research suggests that the effects of other components, such as S and water, cannot be disregarded, and that more complex melting regimes have to be considered (see later), but Gast's interpretations are close to the truth.

---

[2] Light rare earth elements, as opposed to HREE, the heavier ones.

(a)

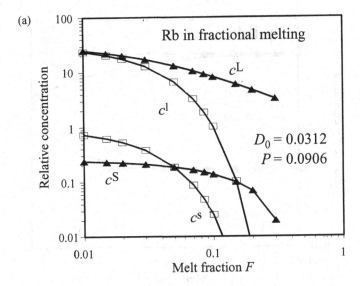

(b)

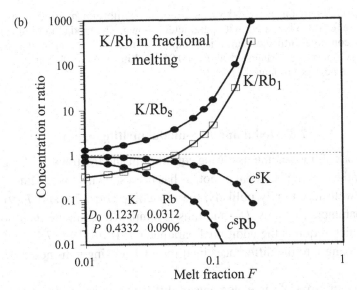

Fig. 7.8 (a) The contrasting behaviour of strongly incompatible Rb during fractional partial melting of ultramafic mantle (Table 7.3). The melting stage ends when $F = 0.34$. (b) Potassium is less incompatible than Rb and so, although the two elements are geochemically very similar, the residual solid retains a higher K concentration during substantial fractional melting than of Rb. Consequently, the ratio K/Rb increases dramatically, both in melt and in residual solid, after the first few % of fractional melting.

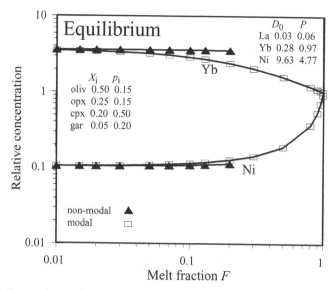

Fig. 7.9 Comparison of modal vs. non-modal equilibrium melting of a garnet peridotite for La, Yb and Ni, with parameters shown. The modal melting concentrations continue until the source is completely molten, but non-modal melting ends when garnet is exhausted at $F = 0.25$; melting would of course continue with some new assemblage.

7.8 Modal and non-modal melting again

The behaviour of trace elements during equilibrium and fractional melting depends on whether the melting is modal or not, as has been shown above. Although modal melting is well-nigh totally unlikely, the modelling (Eqs. 7.9 and 7.19) requires only one parameter, namely D_0, and thus is simpler to compute than non-modal melting, which requires the additional parameter P (Eqs. 7.11, 7.12 and 7.22). It is worth looking into the differences incurred by choosing non-modal over modal formulae.

The concentrations of three elements of different compatibilities have been calculated for different degrees of equilibrium and fractional melting of a garnet peridotite, for modal and for non-modal melting, using appropriate mineral–melt partition coefficients (from Table 7.1) and the mineral proportions shown on the figures. Equilibrium melting curves (Fig. 7.9) show a slightly incompatible Yb and a strongly compatible Ni; the strongly compatible La is not included because the modal and non-modal concentrations were virtually identical. The differences between modal and non-modal concentrations are expressed in percentages in Fig. 7.10a, and are seen to be negligible for La and Ni up to at least 10% melting; Yb, on the other hand, shows substantial differences for as little as 6–7% melting.

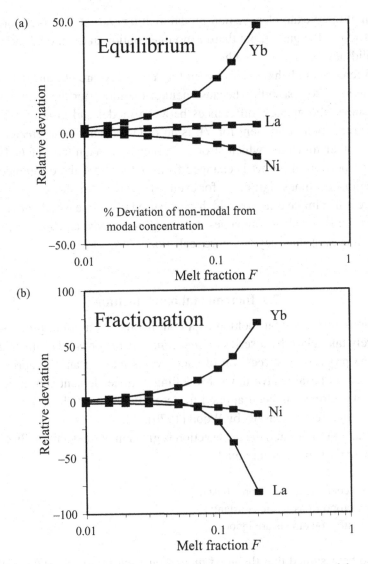

Fig. 7.10 Concentration differences between modal and non-modal melting expressed as % of the modal concentration. (a) In equilibrium melting the greatest differences occur with Yb. (b) In fractional melting the differences are larger, and commence at a lower degree of melting. La shows much larger differences when the degree of melting exceeds 7%, because of its incompatible nature.

Fractional melting behaviour with the same parameters (Fig. 7.10b) shows similar behaviour for Ni, but the deviations are larger for Yb. Lanthanum's incompatibility causes its concentrations to drop off quickly as melting proceeds past 10%, giving significant differences between modal and non-modal melting.

In both of these examples the non-modal melting ends at $F = 0.25$, on account of exhaustion of the garnet, but there is no interruption in the modal melting until all the solids are consumed together.

The differences in behaviour between La, Yb and Ni are not directly functions of the degree of compatibility, because neither the most nor the least compatible element shows the greatest differences between modal and non-modal melting. But the garnet–melt coefficient for Yb is an order of magnitude greater than for any of the other minerals, which strongly affects the magnitude of P. In fact, if the melt proportion of garnet is changed from 0.2 to 0.5% the difference in melt concentrations becomes significant for even 1% degree of melting.

So there is no simple criterion which can be used to choose whether to ignore the greater precision which accompanies the use of non-modal calculations. It should be added, however, that they are more realistic.

7.9 Incremental batch melting

The previous sections treat melting as a continuous or on-going process. It may alternatively take place by a series of *steps, increments* or *batches*, interrupted by halts in the process. A succession of such events may be an appropriate model for sequences of basaltic lava-flows. Modelling of trace element concentrations by incremental melting has been applied to ocean floor basalts by Langmuir *et al.* (1977) and to terrestrial basalts by Wood (1979).

The treatment in this and the next section is given in more detail in Shaw (2000). The following assumptions are made:

(i) partition coefficients D^{i-m} are constant;
(ii) melting proportions p_i are constant;
(iii) kinetic diffusion effects are ignored.

It will also be assumed that the *mass melted on each step* is constant. If the melt fraction on the first step is F_1 then

$$L_1 = F_1 W_0 \tag{7.28}$$

and since $L_1 = L_2 = \cdots = L_n$

$$F_2 = \frac{F_1}{1 - F_1} \quad F_3 = \frac{F_2}{1 - F_2} = \frac{F_1}{1 - 2F_1} \tag{7.29}$$

$$F_n = \frac{F_1}{1 - (n - 1)F_1} \tag{7.30}$$

It may also be noted that since each melt increment is equal to $F_1 W_0$, the total degree of melting f_n after n steps is given by

$$f_n = n F_1 \tag{7.31}$$

It is then easy to show that, since the mass balance for mineral i is, from Eq. (7.5)

$$X_i^1 = \frac{X_i^0 - p_i F_1}{1 - F_1} \tag{7.32}$$

$$X_i^2 = \frac{X_i^1 (1 - F_1) - p_i F_1}{1 - 2F_1} = \frac{X_i^0 - 2 p_i F_1}{1 - 2F_1} \tag{7.33}$$

and

$$X_i^n = \frac{X_i^0 - n p_i F_1}{1 - n F_1} \tag{7.34}$$

Thus, from Eq. (7.7)

$$D_2 = \sum X_i^2 D^{i-1} = \frac{D_0 - 2P F_1}{1 - 2F_1} \tag{7.35}$$

and this may be generalised to give

$$D_n = \frac{D_0 - n P F_1}{1 - n F_1} \tag{7.36}$$

If the trace element concentrations are c_0 and c_1^S in the source and in the solid at the end of step 1, then the trace element mass balance, for an equilibrium process, states that

$$c_0 W_0 = c_1^S W_1^S + c_1^L L_1 = c_1^L [D_1(1 - F_1) + F_1] W_0 \tag{7.37}$$

$$c_1^L = \frac{c_0}{D_1 W_1^S + L_1} \tag{7.38}$$

where D_1 is defined by Eq. (7.36). This may be extended to give

$$c_n^L = c_{n-1}^S \cdot \frac{1}{D_n(1 - F_n) + F_n} \tag{7.39}$$

The concentrations calculated using Eqs. (7.36) and (7.39) for a strongly incompatible element (Fig. 7.11) show that although each step is an equilibrium melting episode, the sequential result is that the liquid concentrations do not tend towards unity, as they would under true equilibrium, but decrease continually (by a straight-line logarithmic trend). The process is thus one of fractionation and, if the mass fractions are very small, the result becomes indistinguishable from Rayleigh fractionation similar to the example in Fig. 7.7b.

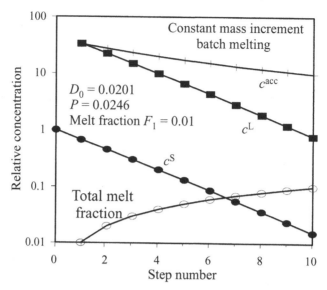

Fig. 7.11 Batch melting with a constant mass increment of a garnet peridotite, with an initial melt fraction of $F_1 = 0.01$, with partition coefficients for La as given by Wood (1979). It is assumed that each step melts in equilibrium, but does not remain at the seat of melting, so that the La liquid log concentration decreases steadily, and the overall effect is of fractionation. After ten steps the total mass melted is 0.10, but the fraction melted on the last step is 0.0109 (see Eq. 7.30).

It should be noted that this process will end when any mineral is totally consumed, which occurs (from Eq. 7.34) when

$$n = \frac{X_i^0}{p_i F_1} \tag{7.40}$$

An alternative, but less useful, formulation assumes a constant melt fraction (F) on each step. It can then be demonstrated that

$$f_n = 1 - (1 - F)^n \tag{7.41}$$

where f_n is the total degree of melting, and consequently

$$D_n = \frac{D_{n-1} - PF}{1 - F} \tag{7.42}$$

and

$$c_n^L = \frac{c_{n-1}^S}{D_n(1 - F) + F} \tag{7.43}$$

With the assumption that F is constant on each step, of course, there is no limit to the possible number of steps, with an ever-diminishing residue.

7.10 Batch melting with retained melt

A more realistic incremental melt model will allow for some of the new melt to remain trapped within the residual solid, and this has been developed by the authors mentioned in the last section. In addition to the assumptions in the previous section, it will be assumed here[3] that

(i) melting only affects the solids present;
(ii) trapped melt is not treated as a phase participating in the melting;
(iii) trapped melt mixes and equilibrates with new melt on the next step.

Consequently, the controlling mass balance is expressed by

$$W_n^S = (1 - F_n)W_{n-1}^S \qquad (7.44)$$

where W_n^S is the mass of residual solid after step n. On each step a mass of melt l_n is produced, and so

$$W_n^S = W_{n-1}^S - l_n \qquad (7.45)$$

If the melt increment is constant then $l_1 = l_2 = \cdots = l_n$, and, as in Eqs. (7.29) and (7.30)

$$F_2 = \frac{F_1}{1 - F_1} \qquad F_n = \frac{F_1}{1 - (n-1)F_1}$$

The total degree of melting f_n after n steps is again given by

$$f_n = nF_1 \qquad (7.46)$$

When the rock mass W_0 begins to melt, the liquid accumulates interstitially without expulsion. When the permeability threshold is reached, a fraction $(1 - Q_1)$ of the melt mass $l_1(=F_1 W_0)$, of mass L_1^{exp} begins to escape, leaving melt mass L_1^r (see Fig. 7.12), and ending step 1

$$L_1^r = Q_1 F_1 W_0 \qquad (7.47)$$
$$L_1^{exp} = (1 - Q_1)F_1 W_0 \qquad (7.48)$$
$$W_1^r = (1 - F_1 + QF_1)W_0 = R_1 W_0 \qquad (7.49)$$

where $R_1 = 1 - F_1 + Q_1 F_1$. The residue, W_1^r, is liquid plus solid.

If the trace element concentrations are c_0 and c_1^S in the source and in the solid at the end of step 1, then the trace element mass balance, for an equilibrium process,

---

[3] An alternative to these three assumptions would be to treat the melt as a melting phase with partition coefficient of unity (e.g. see Albarède, 1995, p. 491 f.).

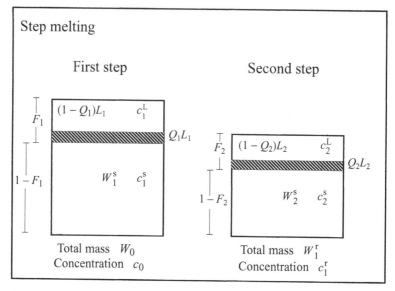

Fig. 7.12 Melting with retention of liquid. A mass W_0 is melted step by step. On the first step a fraction F_1 of melt is formed, from which a fraction $(1 - Q_1)$ is expelled as a lava-flow or intrusion. The residual fraction $Q_1 F_1$, together with the unmelted solids W_1^S, form the residue W_1^r, and the melting continues to the second step, the new melt $F_2 W_1^S$ mixing with the residual melt from step 1.

states that

$$c_0 W_0 = c_1^S W_1^S + c_1^L l_1 = c_1^L [D_1 (1 - F_1) + F_1] W_0 \qquad (7.50)$$

$$c_1^L = \frac{c_0}{D_1 W_1^S + l_1} \qquad (7.51)$$

where D_1 is defined by Eq. (7.36). The concentration in the total residue from step 1 is c_1^r and is obtained from a second mass balance

$$c_1^r W_1^r = c_1^S W_1^S + c_1^L L_1^r \qquad (7.52)$$

$$c_1^r = c_1^L \cdot \frac{D_1 W_1^S + L_1^r}{W_1^r} \qquad (7.53)$$

As depicted in Fig. 7.12, W_1^S now melts to give a liquid l_2, which mixes and equilibrates with the melt residue from step 1, to form a mass L_2, expelling a mass $(1 - Q_2)L_2$ and retaining a mass $Q_2 L_2$, so that by the end of step 2 the mass balance is as follows

$$l_2 = F_2 W_1^S = F_1 W_0 \quad W_2^S = (1 - F_2) W_1^S = (1 - 2F_1) W_0 \qquad (7.54)$$

$$L_2 = l_2 + L_1^r \quad L_2^{exp} = (1 - Q_2)L_2 \quad L_2^r = Q_2 L_2 \qquad (7.55)$$

$$W_2^r = W_2^S + L_2^r = R_2 W_1^S + Q_1 Q_2 F_1 \qquad (7.56)$$

The mass balance for the trace element, before expulsion of fraction $(1 - Q_2)$ of the melt, may be stated as

$$c_1^r W_1^r = c_2^S W_2^S + c_2^L L_2 \tag{7.57}$$

whence

$$c_2^L = c_1^r \cdot \frac{W_1^r}{D_2 W_2^S + L_2} \tag{7.58}$$

For the residue from step 2, the mass balance is written as follows

$$c_2^r W_2^r = c_2^S W_2^S + c_2^L L_2^r \tag{7.59}$$

$$c_2^r = c_1^r \cdot \frac{W_1^r}{W_2^r} \cdot \frac{D_2 W_2^S + L_2^r}{D_2 W_2^S + L_2} \tag{7.60}$$

The approach used for step 2 is readily adapted to subsequent steps. Thus, the mass of new melt produced in step 3 is l_3 and the relevant mass fractions are

$$l_3 = F_3 W_2^S = F_1 W_0 \quad W_3^S = (1 - 3F_1)W_0 \quad L_3 = l_3 + L_2^r \tag{7.61}$$

$$L_3^{exp} = (1 - Q_3)L_3 \quad L_3^r = Q_3 L_3 \tag{7.62}$$

$$W_3^r = W_3^S + L_3^r \tag{7.63}$$

The bulk partition coefficient has already been defined so the trace element melt concentration can be found as follows

$$c_2^r W_2^r = c_3^S W_3^S + c_3^L L_3 = c_3^L \left(D_3 W_3^S + L_3 \right) \tag{7.64}$$

$$c_3^L = c_2^r \cdot \frac{W_2^r}{D_3 W_3^S + L_3} \tag{7.65}$$

The residue concentration may then be calculated as

$$c_3^r W_3^r = c_3^S W_3^S + c_3^L L_3^r \tag{7.66}$$

$$c_3^r = c_3^L \cdot \frac{D_3 W_3^S + L_3^r}{W_3^r} = c_2^r \cdot \frac{W_2^r}{W_3^r} \cdot \frac{D_3 W_3^S + L_3^r}{D_3 W_3^S + L_3} \tag{7.67}$$

Equations (7.65) and (7.67) are identical in form to Eqs. (7.58) and (7.60). It should be noted that the superscripts relating specifically to the melt concentration are capitalised to indicate equilibrium conditions.

It may be desirable to find the average trace element concentration in all the liquids expelled, assuming that they have accumulated and mixed. If the expelled melt mass after n steps is $\bar{L}_n$ then

$$\bar{L}_n = W_0 - W_n^r \tag{7.68}$$

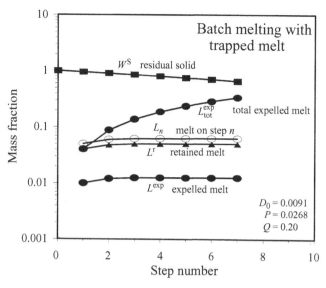

Fig. 7.13 The relative mass fractions shown in Fig. 7.12 depend on the magnitudes of D_0, P, F and Q. The system here is from Table 7.1 and Table 7.4. With a constant mass increment and $F_1 = 0.05$, as chosen here, the curve for W^S is independent of Q, and would go to zero on step 20, when melting is complete. However the values of D_0 and P are such that clinopyroxene is used up by step 7, at which point the total expelled melt mass is 0.3375, and the melting stage ends. The melt mass L_1 on step 1 is independent of Q. Variation in Q will, however, change the configuration of the four evolving melt mass fractions.

so, for example

$$\bar{L}_1 = W_0(1 - R_1) \qquad (7.69)$$

If the accumulated trace element concentration is $\bar{c}_n$ then

$$\bar{c}_n \bar{L}_n = c_0 W_0 - c_n^r W_n^r \qquad (7.70)$$

whence

$$\bar{c}_n = \frac{c_0 W_0 - c_n^r W_n^r}{W_0 - W_n^r} \qquad (7.71)$$

Equations (7.65) and (7.67) may be applied, with appropriate changes of subscript, to subsequent steps. They are more compact than equivalent expressions using the variables F_n and Q_n. If the retention fractions Q are taken as constant, the algebraic expressions become somewhat simpler.

Figure 7.13 shows the sequence of masses generated by successive melt steps in the melting of a spinel lherzolite with the melt rich in clinopyroxene (Table 7.4), using partition coefficients from Table 7.1. There is constant melt mass (L_n) on

Table 7.4 *Model for melting a spinel lherzolite, using selected partition coefficients and mineral proportions*

	Mode X_i^0	Melt p_i	D_{La}^{i-m}	D_{Ti}^{i-m}	D_{Ni}^{i-m}
olivine	0.554	0.2	0.0001	0.3	13.0
orthopyroxene	0.252	0.3	0.0004	0.7	6.6
clinopyroxene	0.173	0.5	0.0536	1.3	4.0
spinel	0.021	0.0	0.0	0.8	16.0
D_0			0.0094	0.59	9.96
P			0.0269	0.92	6.58

Partition coefficients from Table 7.1.

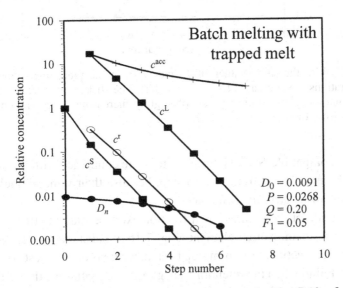

Fig. 7.14 Step melting, with the same parameters as in Fig. 7.13; c^r is from Eq. (7.53). The strongly incompatible element, La, is concentrated in the early melts, but its concentration decreases rapidly.

each step, equal to $F_1 = 5\%$ of the original bulk mass, and 20% of the melt is retained (Q). With this value of F_1, the expelled melt mass would equal the initial solid by step 20 and the melting would be at an end, but in fact the clinopyroxene has all been consumed by step 7, so some other mineral assemblage would take over.

The behaviour of trace elements in the same system is explored in Figs. 7.14, 7.15 and 7.16. The first of these shows that the strongly incompatible element, La, is concentrated in the early melts, and consequently falls off greatly after step 3.

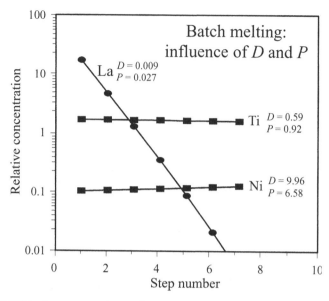

Fig. 7.15 With the same values of F and Q as in the preceding figure, melt concentrations are compared for La, Ni and Ti. The strongly compatible Ni and intermediate Ti show similar trends of slightly increasing concentration, quite different from La.

The strongly compatible Ni and intermediate Ti are contrasted with La in Fig. 7.15. Both show little variation from step to step, because their mineral–melt partition coefficients show no striking differences.

The magnitude of the retained melt fraction exerts a marked effect on the concentration of an incompatible element (Fig. 7.16a); substantial retention reduces any tendency for fractionation from step to step, whereas the greatest concentration contrasts are found when the retention is negligible. By contrast, the concentration trend for a compatible element is independent of melt retention (Fig. 7.16b).

As described in the previous section, an alternative to the constant mass increment is to allow free variation of F_n, which may be appropriate in some circumstances.

7.11 Equilibrium melting vs. fractional melting

Many studies of trace element behaviour in the genesis of basalts focus on the composition of the basalt magmas formed by melting of mantle materials. A study by Johnson *et al.* (1990), however, looked at the composition of residual clinopyroxenes in the deep ocean peridotites from which associated basalts had been derived. Samples were analysed for the REE and other elements from a number of geographically and geophysically distinct locations, and compared with calculated pyroxene

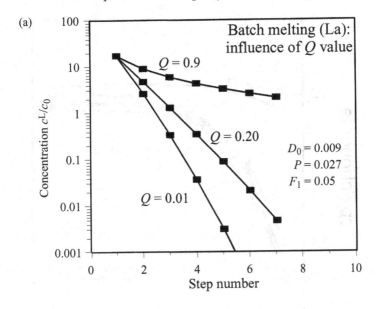

(a)

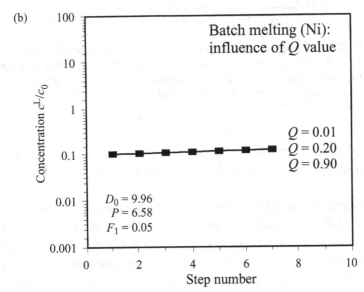

(b)

Fig. 7.16 With the same value of F_1 as in Fig. 7.12, the degree of melt retention influences strongly the concentration trend from step to step of an incompatible element (a), with the least effect when there is very substantial retention. (b) A compatible element's concentration trend is independent of the degree of melt retention.

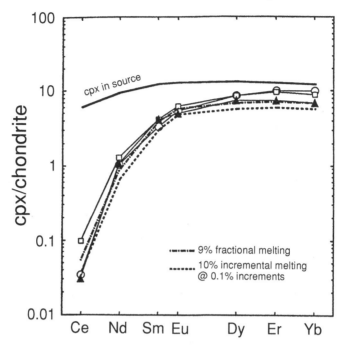

Fig. 7.17 Three clinopyroxenes from the peridotites associated with, and parental to, the Las Islas Orcadas fracture zone basalts, show marked depletion in LREE, relative to chondrites. These data are compared with calculated trends obtained by 9% fractional melting and 10% incremental melting in intervals of 0.1% (after Johnson *et al.*, 1990, Fig. 11a).

from several possible mantle compositions, using both equilibrium (or single-step batch) melting and fractional melting models (i.e. using Eqs. 7.11, 7.12 and 7.23).

In all the samples studied it was found that equilibrium melting was inadequate to account for the REE abundance patterns, which show marked relative depletion of the LREE. Examples of the pyroxenes from the peridotites of the Las Islas Orcadas fracture zone are in Fig. 7.17, and it is seen that these patterns are replicated almost exactly by either 9% fractional melting or by 10% incremental melting made up of 0.1% steps. Similar results coming from the other sample areas make up a convincing case.

7.12 Melting in the presence of volatiles

The effect of volatile components on melting has not been considered explicitly in the preceding sections. It is, however, well-known that the presence of a little water will drastically lower the temperature of the onset of melting of a silicate rock, and

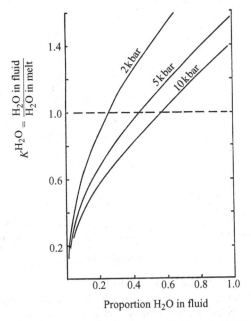

Fig. 7.18 Calculated partition coefficient (after Holloway, 1976, Fig. 4) for H_2O between a binary H_2O–CO_2 fluid and a silicate melt under three confining pressures. With a high proportion of H_2O in the fluid (RHS) the coefficient exceeds unity at all three pressures, which means that there is less H_2O in the melt. With low H_2O in the fluid (high CO_2) the coefficient is < 1 and more H_2O goes into the melt.

may alter the assemblage of minerals on the solidus; both will, in turn, affect trace element behaviour.

The effect of volatile components is known best for H_2O and CO_2, which in nature are usually the most abundant species, but it should be kept in mind that acids, halogens, S compounds and many other compounds will also be present. In the following *fluid, vapour* and *gas* will be taken as synonymous, and distinguished from *melt*.

The minerals of the source rock may contain H_2O and CO_2 which will be released on melting. Small amounts of H_2O and CO_2 will dissolve in a melt but, as the concentration increases, eventually they will be released to form a separate liquid or vapour phase, depending on whether the pressure and temperature exceed the critical point. The possible relationships between H_2O and CO_2 are quite complex and have been modelled by Holloway (1976). In an ideal mixture of the two, partitioned between melt and fluid phases, Fig. 7.18 shows the partitioning of H_2O between melt and fluid is strongly influenced by the CO_2 abundance. For instance, when the fluid phase is nearly all H_2O (RHS), the partition coefficient exceeds unity and so the concentration of water in the melt is less; but when the

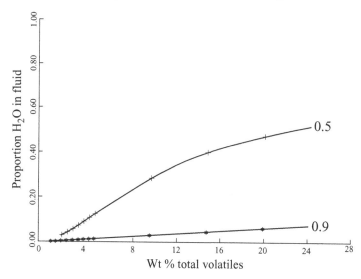

Fig. 7.19 In addition to the ratio of H_2O to CO_2, the partition of either between fluid and melt depends also on the *total* % volatiles. Here is shown the proportion of H_2O in the fluid phase vs. the total system concentration of volatiles, for two values of the ratio of moles CO_2 to total moles H_2O plus CO_2, under a fixed pressure of 10 kbar (after Holloway, 1976, Fig. 5).

fluid is rich in CO_2 and poor in H_2O (LHS), the partition coefficient is less than unity and the melt is richer in H_2O. The effect of total pressure, moreover, is to depress the partition coefficient and so increase the proportion of H_2O going into the melt.

Holloway's study also shows that the proportion of total volatiles (i.e. H_2O plus CO_2, relative to the melt mass) also exerts a control on the partition of water between fluid and melt. In Fig. 7.19 the proportion of H_2O is seen as a function of the initial ratio of CO_2 to total volatiles, and the % volatiles in the system. The lines of constant mole fraction CO_2 trace out the paths followed during isobaric melting as the melt fraction increases.

The behaviour of a trace element will be influenced by the masses of melt (W^m) and volatiles (W^f) and by the partition coefficients mineral–melt (D^{i-m}) and mineral–fluid (D^{i-f}). First it will be useful to look at the possible relationships between fluid and melt, assuming that the initial system is a rock mass containing some volatiles, either surface-adsorbed or as components of the minerals.

The range of possibilities is indicated in Fig. 7.20, which synthesises and simplifies many melting experiments on silicate rocks by P. J. Wyllie and colleagues (2000). The components are shown as pseudo-binary mixtures of 'solid' and 'volatile' components, and possible phases are rock, minerals, liquid (i.e. silicate melt) and vapour. Two diagrams are necessary, to distinguish between sub- and super-critical conditions. On the LHS of each diagram is a field where essentially

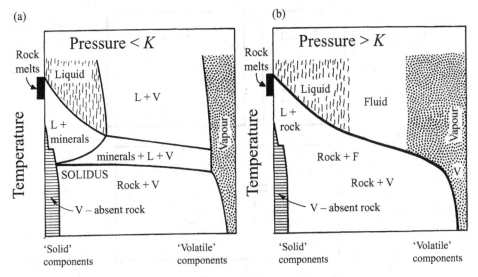

Fig. 7.20 A synthesis and simplification of many individual experiments of melting silicate and carbonate rocks, intended in particular to illustrate melting in the system peridotite–H_2O–CO_2, after Wyllie and Ryabchikov (2000). The components are shown as pseudo-binary mixtures (wt %) of 'solid' and 'volatile' components, and possible phases are rock, minerals, liquid (i.e. silicate melt) and vapour. Two diagrams, i.e. (a) and (b), are necessary to distinguish between sub- and super-critical conditions; above the critical point there is no distinction between liquid and vapour, except in density.

dry melting takes place over a high temperature interval from 'dry' rock; the field of vapour-absent rock, however, includes rocks which contain hydrated minerals or carbonates.

As the volatile content increases, the liquidus temperature is depressed and a separate gas phase becomes stable. In the main part of each diagram the low temperature field shows rock existing with vapour. As heat is supplied, with temperature rise or pressure release, a solidus is intersected and the rock begins to melt. Below the critical point (a) residual minerals occur with liquid and vapour until, as more heat is supplied, the minerals complete their melting at the liquidus, leaving melt and vapour. Above the critical point (b) the distinction between liquid and vapour disappears and, as the authors state 'there is no defined solidus where melting begins' (Wyllie and Ryabchikov, 2000, p. 1200).

In order to construct a model suitable for following element abundances the volatile content can be considered under three possibilities:[4]

(i) fluid mass is just sufficient to saturate the melt: this case corresponds to the LHS of Fig. 7.20a or b, and does not need examination, since there is no separate fluid phase;

---

[4] The following has been modified from the approach used earlier (Shaw, 1978).

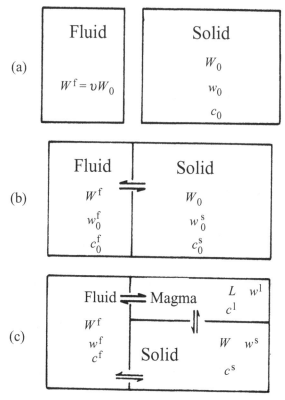

Fig. 7.21 Schematic model of a possible fluid melting process. The rock mass W_0 contains a mass w_0 of the trace element, at concentration c_0. The fluid mass $W^f = vW_0$ contains none of the trace element initially (a), but then equilibrates with the rock (b), generating concentrations c_0^f and c_0^s respectively. During partial fractional melting, the minerals, melt and fluid are in instantaneous equilibrium (c), and the concentrations become c^f and c^s, while the accumulated concentration in the melt is $\bar{c}^l$.

(ii) fluid migrates from the surroundings to the melting site: this is an open system and will not be considered here;

(iii) fluid mass is a fixed proportion (v) of the rock mass: this is a closed system in which $W^f = vW_0$; as melting begins some fluid separates and some is used up to keep the melt mass $L \ (= FW_0)$ saturated, but the fluid remains present as a separate phase.

Both (ii) and (iii) correspond to the central parts of Fig. 7.20, and the following discussion will be based on (iii).

The essentials of the model are shown in Fig. 7.21, using definitions already given. The fluid is here assumed to contain none of the trace element before melting starts, but the following can readily be adapted for an initial trace element fluid mass of w_0^f. The mass balance for the equilibration of fluid and solid phases is given by

$$c_0 W_0 = c_0^f v W_0 + c_0^s W_0 \tag{7.72}$$

So, using Eq. (7.7)

$$\frac{c_0^s}{c_0^f} = D_0^{s-f} = \sum_i X_0^i D^{i-f} \tag{7.73}$$

where D_0^{s-f} is the element's bulk partition coefficient between rock and fluid

$$\frac{c_0^f}{c_0} = \frac{1}{D_0^{s-f} + v} \tag{7.74}$$

and

$$\frac{c_0^s}{c_0} = \frac{D_0^{s-f}}{D_0^{s-f} + v} \tag{7.75}$$

In deriving the equations for trace element behaviour, it will be assumed that fluid and solid are in equilibrium with the melt as it forms (indicated by the arrows in Fig. 7.21; this situation may persist (equilibrium melting) or may not (Rayleigh melting). Assuming non-modal melting, Eq. (7.7) gives

$$D^{s-m} = \frac{D_0^{s-m} - P^{s-m} F}{1 - F} \quad D^{s-f} = \frac{D_0^{s-f} - P^{s-f} F}{1 - F} \tag{7.76}$$

and so

$$D^{f-m} = \frac{D_0^{s-m} - P^{s-m} F}{D_0^{s-f} - P^{s-f} F} \tag{7.77}$$

If it is assumed that p_i, D^{i-m}, D^{i-f} are constant then D^{f-m} also is constant.
It is also useful to note that

$$P^{s-f} = P^{s-m} D^{m-f} \tag{7.78}$$

For the equilibrium case the mass balance is given by

$$c^L F = c_0 - c^s (1 - F) - c^f v \tag{7.79}$$

whence, using Eq. (7.74)

$$c^L = \frac{c_0}{D_0^{s-m} + F(1 - P^{s-m}) + v D^{f-m}} \tag{7.80}$$

Similarly, the other two concentrations are given by

$$c^S = c^L D^{s-m} \qquad c^f = c^L D^{f-m} \tag{7.81}$$

As before, the melting interval will end when some phase i is exhausted, after a fraction F_{end} has melted, i.e. when

$$F_{end} = \frac{X_0^i}{p_i} \tag{7.82}$$

For Rayleigh melting the mass balance is controlled by the differential

$$dw^s + dw^l + dw^f = 0 \tag{7.83}$$

Substitution (see Shaw 1978, Appendix) and integration, as before, gives

$$c^f = \frac{c_0}{D_0^{s-f} + v} \left(1 - \frac{F P^{s-f}}{D_0^{s-f} + v} \right)^{\left[\frac{1}{p^{s-m}} - 1 \right]} \tag{7.84}$$

$$c^s = D^{s-f} c^f \qquad c^l = \frac{c^f}{D^{f-m}} \tag{7.85}$$

The concentration $\bar{c}^l$ in all the accumulated fractions of melt, up to a degree of melting F is given by rearranging the mass balance equation

$$\bar{c}^l F = c_0 - c^s(1 - F) - c^f v \tag{7.86}$$

to obtain

$$\bar{c}^l = \frac{c_0}{F} \left(1 - \left(1 - \frac{F P^{s-f}}{D_0^{s-f} + v} \right)^{\frac{1}{p^{s-1}}} \right) \tag{7.87}$$

As an application of these equations, the melting of a diorite in the presence of excess H_2O can be modelled. It may first be assumed that the fluid phase is in fixed proportion v to the mass melting, when Eqs. (7.74) and (7.75) may be used to follow the effect of v on the initial concentrations. Using the parameters given in Table 7.5 it is seen that Rb will be strongly partitioned into the fluid, relative to the rock (Fig. 7.22a), whatever the proportion of fluid. But, if the fluid mass were to be as great as the rock source (e.g. if $v = 1$), it would become the main reservoir, leaving very little in the rock before melting started. By contrast Ce does not have so strong a preference for the fluid phase (Fig. 7.22b), and the ratio Rb/Ce is nearly constant when the fluid is in great excess. The basic reason for this behaviour lies in the markedly different values of D^{f-m} for Rb and Ce used in Table 7.5.

Looking now at the behaviour of the same elements after melting has started, using Eq. (7.84), it is seen (Fig. 7.23) that the Rb/Ce ratio in the melt is quite

Table 7.5 *Data for partial melting of diorite with fluid phase*

	X_i^0	P_i	$D_{Rb}^{i-m\ a}$	$D_{Ce}^{i-m\ a}$
quartz	0.05	0.3	0.0	0.0
orthoclase	0.10	0.3	0.5	0.044
plagioclase	0.85	0.3	0.03	0.27
D_0^{s-m}			0.0755	0.2339
P^{s-m}			0.159	0.0942
$D^{f-m\ b}$			2.5	1.0
D_0^{s-f}			0.0302	0.2339
P^{s-f}			0.0636	0.0942

[a] D^{i-m} from Shaw (1978) Table 1.
[b] D^{f-m} from Webster *et al.* (1989) Table 4 for 2 kbar and 850°C.

constant at low fractions of melt, but can change markedly as 10% or more melt develops, even with a low fluid fraction.

Although the partition coefficients in Table 7.5 have been arbitrarily chosen, it is clear that different degrees of attraction to the fluid phase can markedly affect the relative behaviour of incompatible trace elements.

Instead of the example just used, a more realistic model could proceed in the following steps:

(i) melting of a rock source containing H_2O and/or CO_2, either in porosity or as mineral components; the released volatiles dissolve in the melt[5] and their concentration decreases as more rock melts;
(ii) aqueous fluids in the surrounding country rocks are attracted to, and dissolve into, the melt bringing some concentration of the trace element of interest; depending on the element its concentration may thus increase as melting proceeds;
(iii) pressure decrease in magma rising by density difference will produce a separation of fluid by retrograde boiling or vesiculation; for some elements there will be a marked discontinuity in melt concentration, accompanied by cooling and crystallisation of the melt.

Such a process sequence can be modelled using equations in this and previous chapters, and has for example been used by Lentz *et al.* (2001) to study magma evolution in Martian shergottites and nakhlites. This topic will be taken up again in Chapter 8.

---

[5] If there is an excess of CO_2 there will be a separate fluid phase from the beginning (Holloway, 1976) which will lead to a different evolution.

(a)

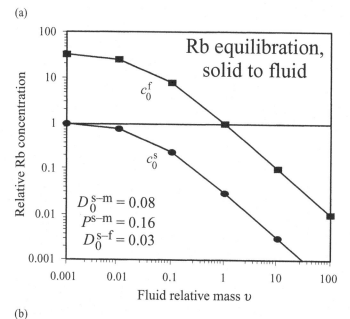

(b)

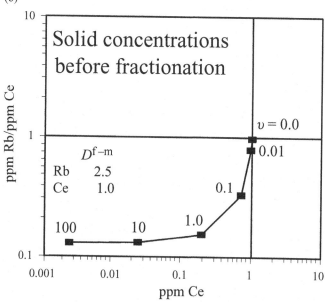

Fig. 7.22 (a) Equilibration between a rock mass having a Rb concentration c_0 and a fluid phase containing no Rb, but of relative mass v, distributes the element according to Eqs. (7.74) and (7.75) and the parameters in Table 7.5. The Rb is strongly enriched in the fluid phase as a function of its relative mass. (b) Rb has a stronger tendency to be enriched in the fluid than Ce, so the initial equilibration of rock and fluid results in a lower Rb/Ce ratio in the rock mass, towards a constant value, as the mass fraction of fluid increases.

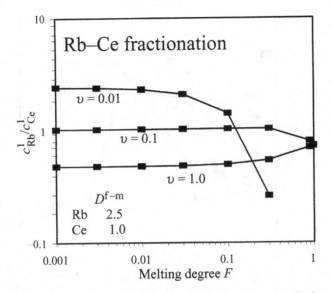

Fig. 7.23 This shows the effect of three values of fluid mass on Rb/Ce fractionation during the course of melting, beginning with the conditions shown in the previous figure. Although the example uses the arbitrarily chosen partition coefficients in Table 7.5, it is clear that different degrees of attraction to the fluid phase can markedly affect the relative behaviour of incompatible trace elements.

7.13 Disequilibrium melting

The preceding discussion has ignored the topic of kinetic effects during melting. In order to achieve either ongoing or momentary equilibrium between trace element concentrations in a mineral and melt it is necessary that the element should diffuse within the solid mineral faster than melting proceeds. The relationship between melting rate and element diffusion rate has been investigated by many authors (Prinzhofer and Allègre, 1985; Bédard, 1989; Bea, 1991; Qin, 1992; Bea, 1996). The last author concludes that mantle rocks require 10^3–10^6 years for melt and residual minerals to equilibrate, although channel flow of melt which entrains xenoliths must attain almost instantaneous upward velocities and, as a consequence, *disequilibrium melting* takes place; this is even more true for high level crustal melting, at the contact between an intruding magma and country rock.

Normally the melt–solid partitioning of an element present as traces in major phases is controlled mainly by the mineral–melt coefficients, D^{i-m}, but also by the mineral proportions in the rock and the liquid, and the degree of melting, F. But if melting is substantially faster than intracrystalline diffusion, then the concentration going into the liquid is the same as in the melting solid and $D_{\text{effective}} \approx 1$

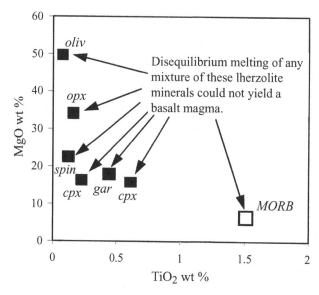

Fig. 7.24 In disequilibrium melting, the minerals of a rock will melt to a liquid of their own composition, and this applies to all the elements they contain. Here are shown typical MgO and TiO_2 concentrations in lherzolite minerals (two cpx values) and in MORB. There is no way to choose proportions of these minerals to match the basalt composition, so it is not possible to explain the lherzolite–basalt relationships by disequilibrium melting.

(Bea, 1996) and the liquid concentration c^l is given by

$$c^l = \sum_i c_i p_i \qquad (7.88)^6$$

where c_i is the concentration in any melting mineral, which makes up a proportion p_i of the melt. The consequences of the disequilibrium melting principle expressed in Eq. (7.88) will now be examined.

Prinzhofer and Allègre (1985) studied the distribution of the REE in harzburgites, residual to melting of the New Caledonian ophiolites to form basalts. After discarding several possible melting processes as ineffective, they proposed that the element distributions could be generated by a disequilibrium melting process, taking place while a garnet–lherzolite source was rising through three depth zones in the mantle, and forming a mid-ocean-ridge basalt or MORB.

The difficulty with disequilibrium melting is that Eq. (7.89) must apply to all the elements in the minerals which melt, including the majors, and it is not possible to melt the minerals of a lherzolite to get a basalt. Figure 7.24 shows the concentrations

---

[6] This equation differs from Eq. (2) of Bea (1996) by omitting the possibility that the process generates new solid phases as well as melt.

of two common oxides, MgO and TiO_2, as reported for peridotite minerals and for MORB, and it is clear that no combination of the minerals can match the basalt composition; *all* contain too much MgO and *none* contain enough TiO_2.

Here then is a paradox. Analysis of melting and diffusion rates leads Bédard (1989) and others to conclude that mantle melting must almost always be a disequilibrium process, but the chemical evidence is not supportive. In addition to the major element problems just described, the terrestrial and lunar rock associations studied by P. W. Gast (see previous section) support equilibrium and fractional mantle melting.

The test presented in Fig. 7.24 can only be used, however, when the source rock and melt differ markedly in mineralogy. It is therefore of no use in studies of quartzo–feldspathic migmatite protoliths which have generated granitoid melts, as discussed in several articles by F. Bea, including the two mentioned above.

The theoretical analysis by Qin (1992) is nevertheless of considerable interest. Qin models the melting of a mineral grain by assuming surface equilibrium with adjacent residual melt, and also assumes that melt is leaving the system or being extracted. The behaviour of a trace element will then lie in a region between the extremes of complete equilibrium melting and complete disequilibrium melting, which Qin calls the *partial equilibrium domain*, according to the magnitudes of the melting rate and the diffusivity of the element within the mineral. In particular, two incompatible elements with identical partition coefficients may be enriched in the melt to different degrees if their diffusivities differ, a process which he terms *diffusive fractionation*. Qin applies this theory to the Pb paradox, which is that the partition coefficients indicate that U is more incompatible than Pb in mantle melting, whereas rock samples indicate the opposite. His solution is that Pb has a higher diffusivity than U and Pb is consequently more incompatible, but there are additional factors operating, as pointed out by Turner *et al.* (2000).

7.14 Accessory minerals entrained during melting

Attention has already been drawn to the fact that the major minerals of a rock commonly contain inclusions of accessory minerals (for example see Fig. 1.8), and also accessory minerals commonly occur near the contacts of major minerals. As a consequence, these accessories will be involved in any melting process. Bea (1996) has looked at the behaviour of trace elements present (both as main and trace components) in them. The elements of concern here include P, REE, Y, Zr, Hf, Th, U, Nb and Ta in oxide-bonded accessories and Ni, Co, PGE[7], As, Sb etc. in sulphides.

---

[7] Platinum group elements.

Lower melting-point accessories will react with melt, going into solution until a saturation level is reached, assuming that the grains are in contact with melt and not isolated as inclusions within major minerals. Watson and Harrison (1984) find that the process depends on three 'fundamental accessory-phase parameters', which are:

(i) the solubilities of the accessory phases;
(ii) the mineral–liquid partition coefficients;
(iii) the diffusivities that govern the rates at which equilibrium is approached.

In the melting of crustal rocks only small amounts of apatite and zircon are needed to saturate the melt, so they do not add large concentrations of P, Zr, Hf and REE to the melt; the remaining apatite and zircon stay in the residue. However, the solubilities of P and Zr are functions of temperature and the bulk melt composition, and some magmas can dissolve large amounts of these minerals.

In more detail, Fig. 7.25 shows (in a modification of Watson and Harrison, 1984, Fig. 1) how mineral solubility may affect the concentration of a trace element melt which is a main component of an accessory mineral, in this case Zr in zircon. The variation of Zr concentration as melting proceeds depends on the abundance of zircon in the rock and its solubility in the melt. If the content of zircon is already sufficient to saturate the melt in Zr (Fig. 7.25a), then the melt concentration remains constant and, as melting reduces the residual mass, its Zr concentration rises. Where the zircon abundance is low, it dissolves until the melt is saturated (Fig. 7.25b), and then at constant concentration the melt mass increases, and more zircon dissolves, until all is consumed; as melting continues the Zr concentration decreases.

A subsequent paper (Watson *et al.*, 1989) investigates the effect of textural structure, in particular whether accessory mineral grains are located within major minerals or at grain boundaries, on melting behaviour. Annealing and partial-melting experiments, as well as theory, show that during high-grade metamorphism and melting, accessory minerals tend to be located at major-phase grain boundaries or in melt, and are larger than included grains, the process being controlled by *interfacial energy minimalisation*. The experiments also showed that, because of their larger size, these grains comprised a very major part of the accessory mineral budget and participated in melting processes, leaving only a small proportion as inclusions, isolated from melting reactions. The results reported by Bea (1996) are very similar.

It is appropriate to conclude with a citation from another paper (Bea, 1995, p. 21): 'The elevated REE fraction contained in accessory minerals . . . [and] the textural position of accessories during anatexis . . . mean that REE-based modelling for deciphering the behaviour of major minerals during melting and crystallisation is of little, if any, use in granitic rocks'.

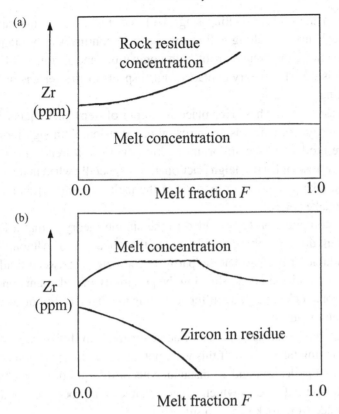

Fig. 7.25 Accessory zircon may behave in one of two ways during melting of its host rock (modified from Watson and Harrison, 1984, Fig. 1). (a) If the content of zircon is already sufficient to saturate the melt in Zr, then the melt concentration remains constant and, as melting reduces the residual mass, its Zr concentration rises. (b) Where the zircon abundance is low, it dissolves until the melt is saturated, and then at constant concentration the melt mass increases, and more zircon dissolves, until all is consumed; as melting continues the Zr concentration decreases.

7.15 Summary

The beginning of melting in a heterogeneous rock is determined by the mineral proportions and may be inferred from an appropriate temperature–composition phase diagram. The melting path will depend on whether equilibrium or fractional melting takes place, but will in any case begin at a eutectic assemblage and temperature. In some systems an anomalous early partial melting takes place, below the eutectic point.

Modelling trace element behaviour during equilibrium and fractional (Rayleigh) partial melting requires knowledge of mineral fractions in the rock and the melt and mineral–melt partition coefficients. In the simplest modelling these parameters are considered as constants. Since the proportions melting differ from the proportions in the rock, a melting stage will end when one of the participating minerals is exhausted; melting will resume, as heat continues to be supplied, with a new assemblage; the behaviour of both compatible and incompatible elements can be contrasted. This theory explains some aspects of the genesis of basalts by mantle melting.

In some cases melting has taken place as a series of steps, increments or batches, each followed by extraction as an intrusion or extrusion. Although individual increments are modelled by equilibrium melting the overall trend in trace element concentrations resembles Rayleigh fractionation, especially when many increments of small size take place. If a proportion of the melt on each step is retained less fractionation takes place.

A further complication to be added to the simple theory is that a fluid phase may be present during melting, in addition to the silicate melt. Although few hard data are available, it is clear that the partition coefficients between fluid and melt vary widely between elements, allowing the possibility that element concentration ratios in igneous rock series may differ according to whether 'dry' or 'wet' melting governed their evolution.

The foregoing theory assumes that trace elements can diffuse within minerals faster than melting takes place. If this were not the case, then partition coefficients would have no application and an instantaneous or disequilibrium melting would take place. In spite of studies which suggest that such a process is possible, there is little evidence from rock series to support it.

The melting of real rocks is complicated by the fact that minerals are not simple pure phases, but commonly carry inclusions of low-melting accessory minerals. In such a case modelling of melting using only the major phases may give misleading results for trace elements concentrated in the accessories, e.g. REE, P or Zr.

References

Allègre, C. G., N. Shimizu and M. Trevill (1977) Comparative chemical history of the Earth, the Moon and parent body of eucrites. *Royal Society of London Philosophical Transactions A*, **285**, 55–67.

Bea, F. (1991) Geochemical modeling of low melt fraction anatexis in a peraluminous system: the Peña Negra complex (central Spain). *Geochimica et cosmochimica acta*, **55**, 1859–74.

(1995) The residence of REE, Y, Th and U in granites and crustal protoliths. Implications for the chemistry of crustal melts. In *The Origin of Granites and Related Rocks*, ed. M. Brown and P. M. Piccoli. United States Geological Survey Circular 1129, pp. 20–1.

(1996) Controls on the trace element chemistry of crustal melts. *Geological Society of American special paper*, **315**, 33–41.

Bedard, J. H. (1989) Disequilibrium mantle melting. *Earth and Planetary Science Letters*, **91**, 359–66.

Doukhan, J.-C. (1995) The phenomenon of early partial melting. *Comptes rendus de l'academie des sciences de Paris*, **321, série Iia**, 1–8.

Gast, P. W. (1968) Trace element fractionation and the origin of tholeiitic and alkaline magma types. *Geochimica et cosmochimica acta*, **32**, 1057–86.

Gurenko, A. A. and M. Chaussiden (1995) Enriched and depleted primitive melts included in olivine from Icelandic tholeiites: origin by continuous melting of a single mantle column. *Geochimica et cosmochimica acta*, **59**, 2905–17.

Holloway, J. R. (1976) Fluids in the evolution of granitic magmas: consequences of finite CO_2 solubility. *Geological Society of America Bulletin*, **87**, 1513–18.

Johnson, K. T., H. J. B. Dick and N. Shimizu (1990) Melting in the oceanic upper mantle: an ion microprobe study of diopsides in abyssal peridotites. *Journal of Geophysical Research*, **95, B3**, 2661–78.

Lange, P. *et al.* (1990) Disorder in vitreous SiO_2, *Journal of Applied Physics*, **68**, 3532–7.

Langmuir, C. H., J. F. Bender, A. E. Bence, G. N. Hanson and S. R. Taylor (1977) Petrogenesis of basalts from the FAMOUS area: Mid-Atlantic Ridge. *Earth and Planetary Science Letters*, **36**, 133–56.

Lentz, R. C. F., H. Y. McSween Jr., J. Ryan and L. R. Riciputi (2001) Water in Martian magmas: clues from light lithophile elements in shergottite and nakhlite pyroxenes. *Geochimica et cosmochimica acta*, **65**, 4551–66.

Luth, W. C., R. H. Jahns and O. F. Tuttle (1964) The granite system at pressures of 4 to 10 kilobars. *Journal of Geophysical Research*, **69, 4**, 759–73.

Mehnert, K. R., W. Büsch and G. Schneider (1973) Initial melting at grain boundaries of quartz and feldspar in gneisses and granulites. *Neues Jahrbuch für Mineralogie, Monatshefte*, **4**, 165–83.

Mysen, B. and I. Kushiro (1976) Melting of peridotite. *Carnegie Institute of Washington Yearbook*, **75**, 546–65.

Ozawa, K. and N. Shimuzu (1995) Open-system melting in the upper mantle. *Journal of Geophysical Research*, **100**, 22315–36.

Presnall, D. C. (1969) The geometric analysis of partial fusion. *American Journal of Science*, **267**, 1178–94.

(1986) An algebraic method for determining equilibrium crystallisation and fusion paths in multicomponent systems. *American Mineralogist*, **71**, 1061–70.

Presnall, D. C. and P. C. Bateman (1973) Fusion relations in the system $NaAlSi_3O_8$–$CaAl_2Si_2O_8$–$KAlSi_3O_8$–SiO_2–H_2O and generation of granitic magmas in the Sierra Nevada batholith. *Geological Society of America Bulletin*, **84**, 3181–202.

Prinzhofer, A. and C. J. Allègre (1985) Residual peridotites and the mechanisms of partial melting. *Earth and Planetary Science Letters*, **74**, 251–65.

Qin, Z. (1992) Disequilibrium partial melting model and its implications for trace element fractionations during mantle melting. *Earth and Planetary Science Letters*, **112**, 75–90.

Rayleigh, Lord J. W. S. (1902) On the distillation of binary mixtures. *Philosophical Magazine, Series 6*, **IV**, 521–37.

Richet, P. and G. Fiquet (1991) High-temperature heat capacity and premelting of minerals in the system MgO–CaO–Al_2O_3–SiO_2. *Journal of Geophysical Research*, **96, B1**, 445–56.

Robertson, J. K. and P. J. Wyllie (1971) Experimental studies on rocks from the Deboullie stock, northern Maine, including melting relations in the water-deficient environment. *Journal of Geology*, **79**, 549–71.

(1978) Trace element behaviour during anatexis in the presence of a fluid phase. *Geochimica et cosmochimica acta*, **42**, 933–43.

(2000) Continuous (dynamic) melting theory revisited. *Canadian Mineralogist*, **38, 5**, 1041–63.

Turner, S., J. Blundy, B. Wood and M. Hole (1984) Accessory minerals and the geochemical evolution of crustal magmatic systems: a summary and prospectus of experimental approaches. *Physics of the Earth and Planetary Interiors*, **35**, 19–30.

(2000) Large ^{230}Th-excesses in basalts produced by partial melting of spinel lherzolite. *Chemical Geology*, **162**, 127–36.

Watson, E. B., E. P. Vicenzi and R. P. Rapp (1989) Inclusion/host relations involving accessory minerals in high-grade metamorphic and anatectic rocks. *Contributions to Mineralogy and Petrology*, **101**, 220–31.

Wood, D. A. (1979) Dynamic partial melting: its application to petrogenesis of basalts erupted in Iceland, the Faeroe Islands, the Isle of Skye (Scotland) and the Troodos Massif (Cyprus). *Geochimica et cosmochimica acta*, **43**, 1031–46.

Wyllie, P. J. and I. D. Ryabchikov (2000) Volatile components, magmas, and critical fluids in upwelling mantle. *Journal of Petrology*, **41**, 1195–2006.

8

Melting: more complex processes

8.1 Introduction

The previous chapter was devoted to the basic aspects of trace element behaviour during melting of a multiphase rock. In assuming that the parameters needed for constructing melting models were constants the simplest equations were obtained. In this chapter some of these assumptions will be relaxed. First to be considered is the case when melting does not occur at a eutectic point, or along a cotectic boundary curve, but where one or more mineral phases melts in a reactive mode. After this the consequences of variable mineral proportions and variable partition coefficients will be taken up, and the last topic will be that of zone refining as a natural process. It will continue to be assumed that minerals and melt react instantaneously, and that rates of element diffusion are much faster than the melting process.

In the last chapter the process of *modal melting* was a useful concept in developing theory. But it is a process which seldom, if ever, can operate, and so, in the remainder of this book, melting will always be taken to be a *non-modal* process.

8.2 Incongruent and reaction melting

When a rock mass begins to melt it may have reached an invariant point in a temperature–concentration phase diagram, i.e. a eutectic, where two or more minerals melt congruently in constant proportions. Alternatively, one of the minerals may melt while reacting to form another mineral, and the temperature is not constant, i.e. it undergoes *incongruent* or *reaction* melting.

The simplest case may be illustrated using the temperature–composition binary phase diagram for forsterite–silica, given previously as Fig. 5.1 and reproduced in a modified manner as Fig. 8.1. If a rock consisting of pure enstatite is subjected to melting, then as the temperature rises nothing happens until T_r is reached, when the enstatite begins to melt incongruently to crystals of forsterite plus liquid. The

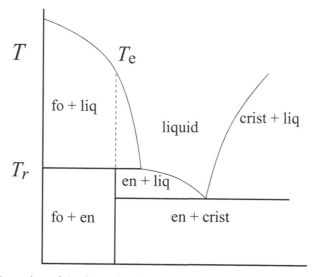

Fig. 8.1 A portion of the forsterite (fo)–silica phase diagram to illustrate the in-congruous melting of enstatite (en). On heating, enstatite begins to melt at tem-perature T_r, forming forsterite. The temperature remains constant until enstatite is consumed, then rises as forsterite reacts with the liquid, whose composition moves up the boundary curve, the proportion of forsterite increasing. When the temperature reaches T_{end} all the forsterite has been consumed.

temperature remains constant until all the enstatite has been consumed, and the system consists of forsterite and liquid. Then the temperature rises, the liquid com-position changes along the liquidus, and forsterite melts, until the temperature reaches T_e. At this point the last forsterite disappears and melting is complete.

If only one mineral melts incongruently the modelling of trace element behaviour is straightforward and will now be examined. Following that, more complex melting relations must be considered.

8.2.1 *Simple incongruent melting*

To understand how a trace element might behave when one mineral melts incongru-ently, we need to consider the mass exchanges, and Fig. 8.2 shows the somewhat more complex system which was used as an example by Hertogen and Gijbels (1976). The first thing to do is find an expression for the bulk partition coefficient D, as follows.

The initial rock mass W_0 consists of minerals a, b, c, d, etc., in proportions X_a^0, X_b^0 etc. Suppose that mineral a melts incongruently, forming melt plus some of each of the other three minerals, i.e.

$$a \rightarrow b + c + d + \text{liquid} \tag{8.1}$$

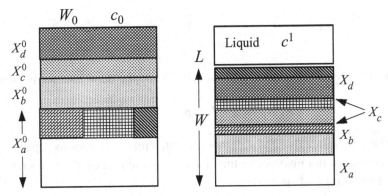

Fig. 8.2 A melting model leading to Eqs. (8.11) to (8.14). A rock composed of four minerals (a, b, c and d), in mass proportions X_i^0 is melting, and mineral a melts incongruently to form proportions t_a, t_b, t_c, t_d and t_1 of the other minerals and of melt, where $\sum t = 1$. After a mass L of melt has formed, the residual solid mass is W and the new mineral proportions are indicated on the right-hand side as X_i. A trace element whose initial concentration in the rock is c_0 has concentration c^l in the melt.

in proportions t_b, t_c, t_d, and t_1, where

$$\sum_i t_i = 1 \tag{8.2}$$

Suppose also that b, c and d are melting congruently at the same time, so it is better to show the reaction as

$$a + b + c + d \rightarrow b + c + d + \text{liquid} \tag{8.3}$$

If the proportion of any mineral phase going into the liquid is p_i then also

$$\sum_i p_i = 1 \tag{8.4}$$

It will be assumed that the parameters t and p are constants. During formation of a fraction F of melt, let a mass M of phase a be consumed, so that the mass balance for mineral a can be written as follows

$$X_a^0 W_0 = X_a W + M \tag{8.5}$$

From the definitions given, then the mass of a going to form liquid is

$$l_a = t_1 \left(X_a^0 W_0 - X_a W \right) = t_1 M \tag{8.6}$$

But also

$$l_a = p_a L \tag{8.7}$$

and so

$$M = \frac{p_a L}{t_1} \tag{8.8}$$

For mineral b the newly formed mass is $t_b M$ and so the mass balance is as follows

$$X_b^0 W_0 = X_b W + p_b L - t_b \cdot \frac{p_a L}{t_1} \tag{8.9}$$

and similarly for minerals c and d. If there are other minerals, such as j, participating in the process, but not involved in the incongruent melting of a then each will also follow an equation such as

$$X_j^0 = X_j + p_j L \tag{8.10}$$

To calculate D we need the mineral proportions after some degree of melting

$$X_a = \frac{X_a^0 - \frac{p_a}{t_1} \cdot F}{1 - F} \tag{8.11}$$

$$X_b = \frac{X_b^0 - \left(p_b - p_a \cdot \frac{t_b}{t_1}\right) \cdot F}{1 - F} \tag{8.12}$$

and similarly for minerals c and d and

$$X_j = \frac{X_j^0 - p_j F}{1 - F} \tag{8.13}$$

We then find that, since $D = \sum X_i D^{i-1}$, then

$$D = \frac{D_0 - QF}{1 - F} \tag{8.14}$$

where

$$Q = \left(\frac{p_a}{t_1}(D^a - D^b t_b - \cdots) + p_b D^b + \cdots + p_j D^j\right) \tag{8.15}$$

and D_0 is calculated as in Eq. (7.7). An alternative formulation for Q is given by Hertogen and Gijbels (1976, Eq. (17)), which may be expressed as follows

$$Q = \left(P + \frac{p_a}{t_1}(D^a - t_1 D^a - t_b D^b - \cdots)\right) \tag{8.16}$$

where P has the same definition as in Eq. (7.7), and the RH term represents the effect of incongruent melting.

To determine trace element behaviour during non-modal fractional melting we now know the partition coefficient and so can apply the basic Rayleigh equation from Chapter 7 (Eq. 7.18)

$$\frac{dc^s}{c^s} = -\left(\frac{1}{D} - 1\right) \cdot \frac{dF}{1 - F} \tag{8.17}$$

and so

$$\frac{c^s}{c_0} = \frac{1}{1 - F}\left(1 - \frac{FQ}{D_0}\right)^{\frac{1}{Q}} \tag{8.18}$$

and

$$\frac{c^l}{c_0} = \frac{1}{D_0}\left(1 - \frac{FQ}{D_0}\right)^{\left(\frac{1}{Q} - 1\right)} \tag{8.19}$$

It is then easy to show that

$$\frac{\bar{c}^l}{c_0} = \frac{1}{F}\left(1 - \left(1 - \frac{FQ}{D_0}\right)^{\frac{1}{Q}}\right) \tag{8.20}$$

It should be noted that F is constrained by three end-of-melting conditions, from Eq. (8.11) to (8.14), namely that

$$F_1 = \frac{X_a^0 t_1}{p_a} \quad F_2 = \frac{X_j^0}{p_j} \quad F_3 = \frac{D_0}{Q} \tag{8.21}$$

The first two relate to exhaustion of particular mineral phases, but F_3 depends on partition coefficients and so varies with the element under consideration.

The corresponding equations for equilibrium or batch non-modal melting are easily derived, giving

$$\frac{c^L}{c_0} = \frac{1}{D_0 + F(1 - Q)} \tag{8.22}$$

As an illustrative example the melting of a garnet peridotite may be considered, in which garnet and clinopyroxene are both melting, but the latter is breaking down incongruently to orthopyroxene and melt.[1] Table 8.1 gives the mineral proportions and partition coefficients, and also includes two choices for the proportion of clinopyroxene going to form melt, and the three criteria for the end of melting (Eq. 8.21). The concentration of an incompatible element (La) in the melt through this incongruent melting reaction (Eq. 8.7) is contrasted with ordinary congruent fractional crystallisation in Fig. 8.3a, where it is seen that when about 10% melting has taken place the concentrations can either be very similar or up to an order of magnitude depressed, depending on the value of t_1.

---

[1] See Shaw (1979).

Table 8.1 *Model for melting garnet lherzolite, using selected partition coefficients and mineral proportions (from Shaw, 1979), with incongruent melting of clinopyroxene to a fraction t_1 of melt and fraction t_{opx} of orthopyroxene*

	Mode X_i^0	Melt p_i	La D_0^i	Yb D_0^i
olivine	0.3		0.008	0.011
orthopyroxene	0.2		0.01	0.15
clinopyroxene	0.3	0.5	0.08	0.4
garnet	0.2	0.5	0.02	10.0
D_0			0.0324	2.1583
P			0.05	0.4
$t_1 = 0.2, t_{opx} = 0.8, Q$			0.19	5.70
$F_1 = X_{cpx}^0/p_{cpx}$			0.12	0.12
$F_2 = X_{gar}^0/p_{gar}$			0.4	0.4
$F_3 = D_0/Q$			0.17	0.38
$t_1 = 0.8, t_{opx} = 0.2, Q$			0.059	5.23
$F_1 = X_{cpx}^0/p_{cpx}$			0.48	0.48
$F_2 = X_{gar}^0/p_{gar}$			0.4	0.4
$F_3 = D_0/Q$			0.55	0.41

For a more compatible element than La, such as Yb, coefficients for which are also in Table 8.1, the effects of the incongruent melting are much less striking. But the relative behaviour of the LREE and HREE is important to basalt genesis, and can be expressed by the ratio La/Yb, which varies very significantly after 10% melting, as seen in Fig. 8.3b.

In these examples it has become clear that the proportion of melt produced by the incongruent melting phase (t_1) is an important parameter, so it must be chosen in conformity with some presumed reaction. In the example just considered, for example, the incongruent reaction could be approximated by the following equation

$$CaMgSi_2O_6 \rightarrow MgSiO_3 + melt \qquad (8.23)$$

formula weight   216   100

By difference, the formula weight of the melt must be 116, so that

$$t_1 = 116/216 = 0.54 \qquad (8.24)$$

and this might be more appropriate than either of the arbitrary choices used above.

In other words the mass balances in melting models need to be based on stoichiometry, and this will be developed in the next section.

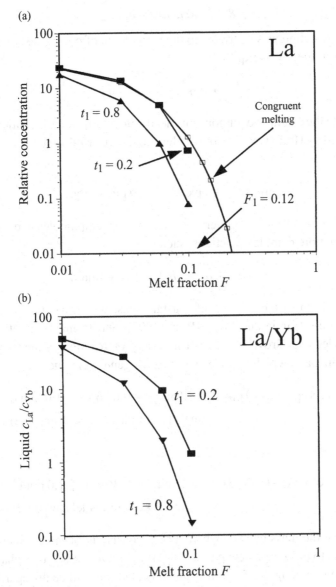

Fig. 8.3 Melting of a garnet peridotite (Table 8.1), with incongruent breakdown of clinopyroxene to melt and orthopyroxene, in proportions t_1 and $(1 - t_1)$; concentrations shown for two values of t_1. The end of melting is controlled by F_1. (a) For the strongly incompatible La, the concentration depends markedly on the proportion of melt formed by the incongruent reaction; (b) Incongruent melting can lead to striking fractionations between elements of different compatibilities, thus the ratio of liquid La/Yb shows order-of-magnitude differences after 10% melting.

8.2.2 Reactive melting

The previous section was restricted to the breakdown of one phase, where mineral *a* melts incongruently as Eq. (8.1)

$$a \rightarrow b + c + d + \text{liquid}$$

In addition to the example of clinopyroxene, the breakdown of pargasitic amphibole has been used by Holloway (1973), Kinzler and Grove (1992) and La Tourette *et al.* (1995)

$$\text{pargasite} \rightarrow \text{cpx} + \text{oliv} + \text{spin} + \text{liquid} \tag{8.25}$$

But, in some circumstances, minerals such as micas, amphiboles and garnets melt according to more complex reactions, such as

$$a_1 + a_2 + a_3 \rightarrow b + c + d + \text{liquid} \tag{8.26}$$

This applies both to the melting of crustal rocks, which has been examined in detail by Benito-Garcia and López-Ruiz (1992), and to mantle melting to form basalt. Kinzler and Grove (1992) list a variety of reactions which they studied experimentally, and which are relevant to basalt genesis, including

$$0.28\,\text{cpx} + 0.19\,\text{opx} + 0.58\,\text{plag} \rightarrow 0.05\,\text{oliv} + 1.0\,\text{liquid}$$

$$\text{(at 9 kbar in plagioclase lherzolite)} \tag{8.27}$$

and

$$1.08\,\text{cpx} + 0.17\,\text{opx} + 0.07\,\text{spin} \rightarrow 0.36\,\text{oliv} + 1.0\,\text{liquid}$$

$$\text{(at 12 kbar in spinel lherzolite)} \tag{8.28}$$

Let us look at the mass balance where a rock consisting of three minerals (a_1, a_2 and a_3) melts to produce a new phase (r) plus melt. Of course other phases b, c or d may be melting congruently at the same time, but this will be disregarded for the present example and considered below.

After a proportion, F, of the original rock has melted the residual solid has mass $W = (1 - F)W_0$ and the liquid mass is $L = FW_0$. There is a mass S of new solid phase r and L of liquid from the reaction. The reaction has consumed a mass M of the reacting phases, a_1, a_2 and a_3 of mass m_1, m_2 and m_3.
Define

t_1 = fraction of M going to form liquid
q_1 = reacting proportion of phase a_1, and similarly for a_2 and a_3

Then

$$M = S + L \qquad (8.29)$$

so

$$M = S + t_1 M \qquad (8.30)$$

or

$$S = (1 - t_1)M \qquad (8.31)$$

Also, each reacting phase contributes liquid and new solid, so

$$m_i = s_i + l_i \qquad (8.32)$$

where

$$\sum m_i = M \quad \sum s_i = S \quad \sum l_i = L \qquad (8.33)$$

From above

$$L = t_1 M \quad m_i = q_i M \qquad (8.34)$$

so

$$m_i = \frac{q_i L}{t_1} \qquad (8.35)$$

and

$$S = F \cdot \frac{1 - t_1}{t_1} \qquad (8.36)$$

Now the proportion of each of the reacting phases can be written, as before

$$X_i W = X_i^0 W_0 - m_i = X_i^0 W_0 - \frac{q_i L}{t_1} \qquad (8.37)$$

so

$$X_i = \frac{X_i^0}{1 - F} - \frac{q_i F}{t_1(1 - F)} \qquad (8.38)$$

The proportion of the new phase is initially zero, so

$$X_r W = S \quad \text{and} \quad X_r = \frac{F(1 - t_1)}{t_1(1 - F)} \qquad (8.39)$$

The partition coefficient can now be written as

$$D = \sum X_j D^j = \frac{D_0 - RF}{1 - F} \qquad (8.40)$$

where

$$D_0 = X^0_{a_1} D^{a_1} + X^0_{a_2} D^{a_2} + X^0_{a_3} D^{a_3}$$

(8.41)

and

$$R = \sum_{i=a_1}^{a_3} \frac{p_i K^i}{t_1} - \frac{(1 - t_1) K^r}{t_1}$$

(8.42)

It should be noted that there will be terminal values

$$F_1 = \frac{t_1 X^0_i}{p_i} \qquad F_2 = \frac{D_0}{R}$$

(8.43)

The expressions (8.42) and (8.43) may now be used to calculate the trace element concentrations, using Eqs. (8.18), (8.20) and (8.22) as follows

$$c^l = \frac{c_0}{D_0} \left[1 - \frac{FR}{D_0} \right]^{(\frac{1}{R} - 1)}$$

(8.44)

$$\bar{c}^l = \frac{c_0}{F} \left[1 - \left(1 - \frac{FR}{D_0} \right)^{\frac{1}{R}} \right]$$

(8.45)

$$c^L = \frac{c_0}{D_0 + F(1 - R)}$$

(8.46)

In each case where a model is to be constructed it is necessary to find a stoichiometric expression for the reaction. That is because the stoichiometry makes it possible to write the mass balance, and calculate the values of t_1, p_i and q_i. For example, the melting of some pelitic rocks can be modelled by the reaction

$$4\,\text{bio} + 8\,\text{sill} + 19\,\text{qtz} \rightarrow 5\,\text{cord} + 4\,\text{liquid}$$

(8.47)

This reaction was proposed by Zeck (1970) and adapted by Benito-Garcia and López-Ruiz (1992) in the study already mentioned. It is essentially the case just examined namely

$$\text{three reacting phases} \rightarrow \text{new phase} + \text{melt}$$

(8.48)

The formula units will be treated as the following species

biotite $H_2KMg_{2.5}Al_2Si_{2.5}O_{12}$ sillimanite Al_2SiO_5
quartz SiO_2 cordierite $Mg_2Al_4Si_5O_{18}$

Using the formula weights (FW) the stoichiometry can be balanced by writing
the reaction as follows

$$4 \text{ bio} + 8 \text{ sill} + 19 \text{ qtz} \rightarrow 5 \text{ cord} + \text{melt}$$

FW	418.2	162.1	60.1	585.1	(1185.6)	
Mass	1673	1297	1142	2926	(1185.6)	(8.49)
Mass Sum			4112		4112	

This equation then defines the reacting proportions q_i of each mineral, in terms of
the total relative mass, the proportion t_1 of melt, and the proportion $(1 - t_1)$ of the
new phase. So

$$q_{\text{bio}} = 1673/4112 \quad q_{\text{sil}} = 1297/4112 \quad q_{\text{qtz}} = 1142/4112 \qquad (8.50)$$
$$= 0.407 \qquad\qquad = 0.315 \qquad\qquad = 0.278$$

$$\sum q_i = 1.0 \qquad\qquad (8.51)$$
$$t_1 = 1186/4112 = 0.288 \ (1 - t_1) = 0.712 \qquad\qquad (8.52)$$

The actual proportion (X_i) of each mineral in the rock is given by Eqs. (8.37), (8.38)
and (8.39) so D can be calculated from Eqs. (8.41) and (8.42).

8.2.3 Three reacting phases plus an inert phase

Consider the reaction discussed in the previous section plus garnet as a present but
inactive phase. We now write the equation as follows

$$1 \text{ gar} + 4 \text{ bio} + 8 \text{ sill} + 19 \text{ qtz} \rightarrow 1 \text{ gar} + 5 \text{ cord} + \text{melt} \qquad (8.53)$$

Assuming one unit of garnet to be present[2] as almandine, $Fe_3Al_2Si_3O_{12}$, then its
formula weight is 498, which is the mass addition to each side of the stoichiometric
equation, whose mass sum is now

$$4112 + 498 = 4610 \qquad\qquad (8.54)$$

so

$$X^0_{\text{gar}} = 498/4610 = 0.108 \qquad\qquad (8.55)$$

The reacting proportions q_i remain the same, because the extra phase does not react,
i.e.

$$q_{\text{bio}} = 1673/4112 = 0.407 \qquad\qquad (8.56)$$

---

[2] The choice of one unit of garnet is arbitrary, and if more is included then X^0_{gar} and the other mineral proportions
will change.

Table 8.2 *Model for reaction melting of biotite–sillimanite–quartz gneiss with selected partition coefficients and mineral proportions, both without and with garnet present, from Benito-Garcia and Lopez-Ruiz (1992)*

	Mode I X_i^0	Mode II X_i^0	q_i	Rb X_i^0	V X_i^0	Yb X_i^0
biotite	0.57	0.51	0.407	2.0	20	1.0
sillimanite	0.25	0.22	0.315	0.1	2	0.01
quartz	0.18	0.16	0.278	0.01	0.008	0.01
cordierite	0.0	0.0	0.0	0.2	0.7	1.5
garnet	0.0	0.11	0.0	0.008	8	20
Total	1.0	1.00	1.00			
t_1	0.2883	0.288				
$A =$				2.9412	30.418	1.4317
$B =$				0.4937	1.728	3.7024
$R =$				2.4476	28.690	−2.2708
D_0^{Rb}	1.1668	1.0445				
D_0^{V}	11.901	11.521				
D_0^{Yb}	0.5743	2.7138				
F_1 - biotite	0.40	0.36				
F_1 - sillimanite	0.23	0.20				
F_1 - quartz	0.19	0.17				

$$A = \sum q_i^* D_i / t_1 \quad B = (1 - t_1)D_{cord}/t_1 \quad R = A - B$$

and t_1 remains unchanged at 0.288. The formulation of X_i for the reacting phases is also unchanged and the inert phase proportion X_j is obtained as follows

$$X_j^0 W_0 = X_j W \tag{8.57}$$

or

$$X_j = \frac{X_j^0}{1 - F} \tag{8.58}$$

But the initial mineral proportions now are different and therefore so is the value of D_0.

An example of this reaction process is given in Table 8.2, where the mineral proportions and partition coefficients for Rb, Yb and V are taken from Benito-Garcia and López-Ruiz (1992) (this treatment is somewhat different from that paper, where there are three preliminary melting steps before the reaction with biotite begins). The partition coefficients used here are not well-established but are sufficient to give some indication of the trends. Mode I is the reaction without garnet and Mode II is with it. The terminal values of F_1, for complete exhaustion of a phase, are given and it is seen that this melting reaction will end with the exhaustion of quartz after about 20% melting, both in the presence and absence of garnet.

Figure 8.4a indicates the variation in the proportions of the minerals, including the growth of cordierite, as melting proceeds. The reacting minerals begin to decrease significantly after a few % of melt has formed, but the garnet abundance increases in compensation. By the time that quartz has been exhausted, at about 20% melt, the abundance of cordierite has grown to 62% of the residual minerals.

The behaviour of Rb and Yb, calculated from Table 8.2, is shown in Fig. 8.4b. Rubidium is behaving incongruently and its melt concentration increases with F; Yb behaves congruently and its concentration decreases when F exceeds 10%. The ratio is dominated by the presence or absence of garnet, because Yb concentrations are lowered by almost an order of magnitude when garnet is present, because of its high partition coefficient. But the ratio in either case increases with F.

8.2.4 More complex reactions

We have encountered four kinds of mineral behaviour in the preceding sections:

  (i) phases which melt congruently;
 (ii) phases which melt incongruently or by reaction;
(iii) phases which are reaction products and are not present in the starting assemblage;
 (iv) phases which are present before and after the reaction, but are unconsumed during melting.[3]

To allow for all possibilities, however, it is necessary also to specify:

  (i) phases which melt congruently but are formed during the reaction e.g. in the example used in Fig. 8.2 phases b, c and d are all melting congruently at the same time as forming from breakdown of phase a;
 (ii) phases which are present initially, are unconsumed, but are also produced by the reaction e.g. melting a pargasitic lherzolite under conditions where olivine is not on the liquidus.

Each of these phases will contribute a term to the value of the bulk partition coefficient D which reflects the behaviour of its proportion X_i. The structure of these terms is summarised in Table 8.3, modified from Benito-Garcia and López-Ruiz (1992).

---

[3] Such phases are considered 'inert' by Benito-Garcia and López-Ruiz (1992) but they can nevertheless affect trace element concentrations.

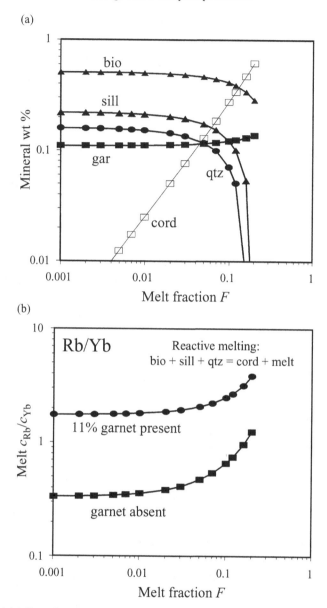

Fig. 8.4 (a) Mineral proportions during the non-modal reaction melting of biotite, sillimanite and quartz to produce cordierite and melt, with garnet present as a non-reacting phase, using data in Table 8.2, in part taken from Benito-Garcia and Lopez-Ruiz (1992). (b) The ratio of the melt concentrations of two trace elements during the reaction melting in the presence and absence of garnet; which does not participate in the reaction, but has a high partition coefficient for Yb.

Table 8.3 *Values of the melting parameters for each kind of mineral phase, during non-modal reaction melting of a multiphase rock (modified from Benito-Garcia and Lopez-Ruiz, 1992, Table 1)*

Type of Phase	p_i	q_i	t_1	X_i^0
Congruent melting	>0	0	0	>0
Incongruent melting	0	>0	>0	>0
New; produced by melting	0	0	>0	0
Congruent; produced by melting	>0	0	>0	0
Unconsumed; inert	0	0	0	>0
Unconsumed; produced by melting	0	0	>0	>0

8.3 Variations in mineral proportions and partition coefficients

In the expressions derived earlier for melting behaviour, it has been taken that the bulk partition coefficient D is a function of the degree of melting F, and of the two parameters D_0 and P. It may be, however, that these two are not parameters, but must be treated as variables because the mineral proportions (X_i) and mineral–melt partition coefficients (D^{i-1}) vary while the melting is taking place.

Variations in X_i and D^{i-1} are controlled by temperature, pressure and chemical composition, and these in turn control the melting behaviour, but the degree of melting F does not lend itself easily to expressing variations in mineralogy and element partition. Nevertheless, as a first approach it will be taken that X_i and D^{i-1} can be expressed in terms of F, and that this can be adapted in a given situation to the real variables. For simplicity it will be assumed that the variability of each is linear with F, so that the mineral proportions in the melt (p_i) and the rock can be represented by the equations

$$p_i = p_i^0 + u_i F \tag{8.59}$$

and the partition coefficients by the equations

$$D^{i-1} = D_0^{i-1} + v_i F \tag{8.60}$$

where u_i and v_i are constants.

The theory needed to model element concentrations has been provided by Hertogen and Gijbels (1976) and it is convenient to start with the bulk partition coefficient of a rock. If we write $D_0 = \sum_i X_i^0 D^{i-m}$ and $P = \sum_i p_i D^{i-m}$ then Eq. (7.7) gives

$$D = \frac{D_0 - PF}{1 - F} = \frac{1}{1 - F}[D_0 + aF - bF^2 - cF^3] \tag{8.61}$$

where

$$D_0 = \sum_i X_i^0 D_0^{i-1} \quad a = \sum_i X_i^0 v_i - \sum p_i^0 D_0^{i-1} \tag{8.62}$$

and

$$b = \sum u_i D_0^{i-1} + \sum p_i^0 v_i \quad c = \sum u_i v_i \tag{8.63}$$

Equation (8.61) may then be substituted into Eq. (7.27), to give

$$\frac{dw^1}{w_0 - w^1} = \frac{dF}{D_0 + aF - bF^2 - cF^3} \tag{8.64}$$

Integration of the LHS of this equation and conversion to concentration gives

$$-\ln \frac{w_0 - w^1}{w_0} = \ln \frac{c_0}{c_0 - \bar{c}^1 F} \tag{8.65}$$

Possible solutions to the integration of the cubic polynomial on the RHS were given by Hertogen and Gijbels (1976, Appendix).

8.3.1 Variation in mineral proportions

In most circumstances variations in the mineral proportions are not important. Rock melting begins at a eutectic, and if the phases participating are simple compounds, their proportions will not change until one of them is completely consumed; if solid solutions the situation will not differ greatly. The possibility for variation in the mineral proportions applies to melting along a cotectic curve, such as the boundary curve between the fields of tridymite and anorthite in Fig. 4.1a. As already discussed in the accompanying text, the ratio anorthite/tridymite at point X is 1.0, but at G it is 0.25, so it would be possible to construct a model relating variation of p_i with F in such a system.

8.3.2 Variation in partition coefficients

If it is decided to neglect the mineral proportion variations then Eq. (8.61) is left with variation in partition coefficients

$$D = \sum X_i D^{i-1} = q + nF + mF^2 \tag{8.66}$$

where

$$q = \sum X_i^0 D_0^{i-1} \quad n = \sum X_i^0 v_i - \sum p_i D_0^{i-1} \quad m = -\sum p_i v_i \tag{8.67}$$

and so

$$\frac{dw^l}{w_0 - w^l} = \frac{dF}{q + nF + mF^2} \tag{8.68}$$

Three solutions to the integration of the RHS of Eq. (8.68) were given by Hertogen and Gijbels (1976) as Appendix Eqs. (A-1), (A-2) and (A-3). The first and last contain mistakes, as was pointed out by Apted and Roy (1981) and the corrected formulae are as follows

Case 1: $n^2 - 4mq > 0$ and so $h = (n^2 - 4mq)^{\frac{1}{2}}$

$$\bar{c}^l = \frac{c_0}{F} \left[1 - \left(\frac{2mF + n + h}{2mf + n - h} \cdot \frac{n - h}{n + h} \right)^{\frac{1}{h}} \right] \tag{8.69}$$

Case 2: $n^2 - 4mq < 0$ and so $k = (4mq - n^2)^{\frac{1}{2}}$

$$\bar{c}^l = \frac{c_0}{F} \left[1 - \exp \frac{2}{k} \left(\tan^{-1} \frac{n}{k} - \tan^{-1} \frac{2mF + n}{k} \right) \right] \tag{8.70}$$

Case 3: $n^2 - 4mq = 0$

$$\bar{c}^l = \frac{c_0}{F} \left[1 - \exp \left(\frac{2}{2mF + h} - \frac{2}{n} \right) \right] \tag{8.71}$$

The concentration in the residual solid is c^s where

$$c^s = \frac{c_0 - \bar{c}^l F}{1 - F} \tag{8.72}$$

The instantaneous concentration c^l is obtained *either* as c^s/D *or* by differentiating the integrated Eq. (8.68), as

$$c^l = \frac{1}{W_0} \frac{dw^l}{dF} = c_0 \frac{d}{dF} (1 - g(F)) \tag{8.73}$$

where $g(F)$ is the function of F in one of the three quadratic solutions just given. In Case 1, for example

$$\frac{c^l}{c_0} = \frac{1}{h} \left(\frac{u}{v} \right)^{(\frac{1}{h} - 1)} \cdot \frac{d}{dF} \left(\frac{u}{v} \right) \tag{8.74}$$

and

$$\frac{d}{dF} \left(\frac{u}{v} \right) = \frac{4mh(n - h)}{(n + h)(2mF + n - h)^2} \tag{8.75}$$

where u/v is the ratio in Eq. (8.69).

The effect of linearly increasing partition coefficients on trace element concentration is seen in Fig. 8.5. The data in Table 8.4 are those used by Hertogen and

Table 8.4 *Model for lherzolite melting with Sc partition coefficients increasing as linear functions of F, as melting progresses (Hertogen and Gijbels, 1976)*

	X_i^0	p_i	D_0^i	u_i
olivine	0.5	0.1	0.1	0.071
orthopyroxene	0.3	0.2	1.0	0.714
clinopyroxene	0.2	0.7	3.0	3.497
Total	1.0	1.0		

$$\sum X_i^0 D_0^i = \quad 0.95 \quad = q$$
$$\sum p_i D_0^i = \quad 2.31$$
$$\sum X_i^0 v_i - \sum p_i D_0^i = \quad -1.3609 \quad = n$$
$$\sum p_i v_i = \quad 2.5978 \quad = -m$$
$$n^2 - 4mq = \quad 11.7237$$
$$\sqrt{(n^2 - 4mq)} = \quad \pm 3.424 \quad = h$$

From Eq. (8.61): $\quad D = (0.95 - 1.3609F - 2.5978\,F^2)/(1 - F)$

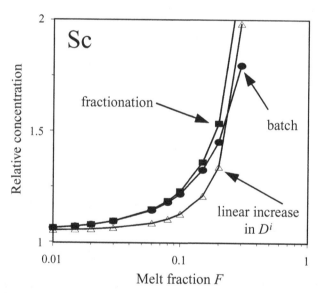

Fig. 8.5 The concentration of Sc during fractional melting of a lherzolite, where partition coefficients are increasing linearly with F, is contrasted with equilibrium and fractional melting with constant coefficients, using data and parameters from Hertogen and Gijbels (1976) in Table 8.4. Variable coefficients initially produce lower concentrations than equilibrium or fractional melting, but this situation is reversed as clinopyroxene is nearly all consumed, at about 29% melting.

Gijbels (1976) to model Sc behaviour during melting of a lherzolite. Their chosen values of D_0^{i-1} and u_i lead to finding that $(n^2 - 4mq) > 0$, so that melt concentrations of Sc have been calculated using Eqs. (8.69) and (8.73). There is little difference between c^l calculated from equilibrium and fractional melting with constant coefficients, and from variable coefficients, at values of a few % melting. But when more than 20% of the solid has melted the position has changed, and the variable coefficient concentration is intermediate between normal equilibrium and fractional melting.

In any case, the differences in concentration shown in Fig. 8.5 are rather insignificant. A considerably greater variation in D^{i-1}, coupled with high quality analyses, would be necessary to permit discrimination between three such model concentration trends.

8.4 Rock melting by zone refining

Zone refining is an industrial process to purify metal rods or cylinders. The rod passes through a furnace which provokes melting in a narrow zone which passes along the rod as it moves. Low-melting components[4] accumulate in the melt and so are progressively swept out, leaving less in the residue. If the rod is made in a continuous loop the process may be continued for several cycles, leaving the metal in high purity. The theory is given by Pfann (1952) and early applications to geochemistry were made by Harris (1957) and Yaroshevsky et al. (1971).

The process may be understood in reference to Figure 8.6. A narrow zone of melting of thickness h is imposed on a metallic rod by an external furnace. The rod is passed slowly from right to left so that the molten zone moves to the right, starting at the origin, melting the rod as it advances; behind the zone the liquid freezes again. If a soluble impurity has a lower melting point than the metal, then its partition coefficient $D < 1$ and the impurity is scavenged out of the metal, accumulating in the liquid. If its concentration in the unmelted rod is c_0, then the concentration in the refrozen rod behind the zone is

$$c^s = Dc^l \qquad (8.76)$$

where c^l is the concentration in the liquid. Now c_0 is the concentration in the liquid at the point of melting, but the overall concentration depends on how far the zone has travelled along the rod, or the number of *zone lengths*, equal to x/h (see Fig. 8.7).

---

[4] Which are called incompatible elements in this book.

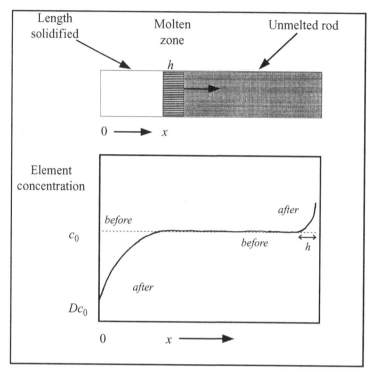

Fig. 8.6 In zone melting a rod of metal is moved laterally, while a narrow zone of melting (width h) is applied by an external furnace. The ratio of the distance x travelled to h measures the number of zones traversed and is called the *zone length*. The leading edge of the zone is melting metal which contains a concentration c_0 of some lower-melting solute; at the trailing edge melt crystallises to a solid containing a concentration Dc_0 of the solute, where D is its partition coefficient. In the case shown here, for which the technique is exploited, D is $\ll 1$, and the metal in the advancing molten zone has become more pure (after Pfann, 1952, Figs. 4 and 5).

Assume that

c_0 is invariant with x, and is in units of g per unit volume

w is the mass of solute in the melting zone, of unit cross-section and thickness h, at any x

w_0 is the mass of solute in the zone at $x = 0$

D is invariant

Then if the zone (at distance x) advances by dx, a volume dx of solid will form containing Dc^ldx of solute; also a volume dx of rod will melt, containing c_0dx of solute. The solute mass within the zone was w, so its mass balance is given by

$$dw = c_0 dx - Dc^l dx \qquad (8.77)$$

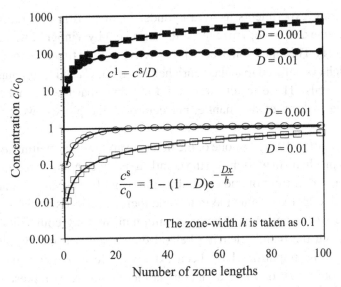

Fig. 8.7 Variation in concentration of a trace element along a rod (c^s) and in the melt (c^l) with the number of zone lengths traversed, during zone melting. The solid concentration becomes asymptotic to c_0, regardless of the magnitude of D, but the relative melt concentration becomes asymptotic to $1/D$. The crystallisation of the final zone is not shown.

or, since

$$c^l = w/h$$

then

$$dw = \left(c_0 - w \cdot \frac{D}{h} \right) dx \qquad (8.78)$$

Using the fact that $w_0 = c_0 h$ and that $c^s = Dc^l = Dw/h$, Eq. (8.78) may be integrated to give the concentration in the crystallised solid as in Pfann's Eq. (2) (1952)

$$\frac{c^s}{c_0} = 1 - (1 - D)e^{-\frac{Dx}{h}} \qquad (8.79)$$

So at the left-hand end, where $x = 0$, the initial value is Dc_0. As the zone moves along the rod c^s increases and, eventually, becomes asymptotic to c_0. The liquid concentration c^l follows a parallel curve offset by a factor $1/D$ (see Fig. 8.7). At the RH end of the rod the value of c^l rises asymptotically to a maximum of c_0/D. If this last liquid zone fractionally crystallises then c^l will rise much higher.

The theory just given was modified by Yaroshevsky (1970) to allow for a variable partition coefficient and variable concentration along the rod.

Zone-melting experiments using a cylinder of meteoritic material, consisting mostly of olivine and orthopyroxene, were carried out by Vinogradov, Yaroshevsky and Ilyin (1971). The repetition of each process five to eight times led to formation of a liquid which crystallised to a glass enriched in Si, Al, Ca and Fe, containing large pyroxene crystals. These results were used to argue that the Earth's crust arose through zone melting of the mantle, but details of the process were not clearly specified.

However, Harris (1957) was one of the first to examine zone melting as a possible petrological mechanism working on the mantle to produce potassic basalts, pointing out that 'solution in front of and crystallization behind an ascending magma body could have an effect very much akin to zone melting' (Harris, 1957, p. 200). The molten zone would be a liquid body of minimum melting composition moving upward through the mantle, leaving behind the crystallised mantle phases from which the soluble impurities have been extracted. He lists many such elements which we would now describe as incompatible, taking K as representative, but also including volatiles such as H_2O and CO_2. The action of zone refining could concentrate K much more effectively than fractional melting, if the process could be understood as acting over many zone lengths with a partition coefficient close to zero.

The dehydration of a crustal slab undergoing subduction releases aqueous fluids which may rehydrate overlying mantle material. The partitioning of incompatible elements between the fluid and an infertile peridotite mantle wedge has been treated as a zone refining process by Ayers (1997). The volatile infiltration scavenges trace elements from the peridotite, carrying them to a seat of melting where island arc basalt magma is forming. The calculated concentrations of ten incompatible trace elements agree well with measured IAB abundances.

It may be concluded that incompatible elements will be strongly scavenged into rising mantle melt, if it can interact with enclosing wall-rocks. But the melt will have to persist without wholesale melting and without solidification, while the zone-melting process proceeds.

8.5 Summary

This chapter considers the effect on trace element concentrations of some melting paths not considered previously.

The manner of mineral melting is governed by heterogeneous-phase reactions, and in some cases the melting is accompanied by conversion to another solid phase. Incongruent or reaction melting of this kind introduces complications in trace-element behaviour, because not only does the partition coefficient for the melting mineral have a role, but also the new mineral being formed by the reaction.

In fact the trace-element concentration in a melt is affected by the partition coefficient of every mineral in the rock, including those which are not melting. There are at least six different kinds of mineral roles, and it is consequently important to have a clear and quantitative model; this usually requires construction of a stoichiometric statement of the reaction process.

In another situation it is possible that both the mineral proportions and their partition coefficients change during the melting process. If this is the case then the controlling variables are temperature, pressure, chemical composition and energy supply, none of which correlate very directly with the degree of melting, F, which has been taken as the main variable in modelling trace element behaviour. If, however, assumptions are taken that mineral proportions and partition coefficients vary in a linear fashion with F, then it is possible to obtain equations which predict the melt concentration.

It appears likely that variation of mineral proportions is less important and, if this factor is disregarded, it is found that partition-coefficient variation can play a major role in controlling the element concentration. Nevertheless it is unlikely that modest partition-coefficient variation would produce concentration variations which could be readily detected.

The zone refining process used in metallurgy may provide a model for some kinds of magma formation. In zone refining a metal rod is passed through a furnace so that a molten zone migrates along it, sweeping out elements of low melting point (low partition coefficient). If magma rises during mantle melting and continues to react with wall-rock, a similar process of enrichment in incompatible and volatile elements will take place.

References

Apted, M. K. and D. S. Roy (1981) Corrections to the trace element fractionation equations of Hertogen and Gijbels (1976). *Geochimica et cosmochimica acta*, **45**, 777–8.

Ayers, J. C. (1997) Trace element modeling of aqueous fluid–peridotite interaction in the mantle wedge of subduction zones. *Seventh Annual V. M. Goldschmidt Conference, LPI Contribution No. 921*, Houston: Lunar and Planetary Institute, p. 9.

Benito-Garcia, R. and J. Lopez-Ruiz (1992) Mineralogical changes of the residual solid and trace element fractionation during partial incongruent melting. *Geochimica et cosmochimica acta*, **56**, 3705–10.

Harris, P. G. (1957) Zone refining and the origin of potassic basalts. *Geochimica et cosmochimica acta*, **12**, 195–208.

Hertogen, J. and R. Gijbels (1976) Calculation of trace element fractionation during partial melting. *Geochimica et cosmochimica acta*, **40**, 313–22.

Holloway, J. R. (1973) The system pargasite–H_2O–CO_2: a model for melting of a hydrous mineral with a mixed-volatile fluid. I. Experimental results to 8 kbar. *Geochimica et cosmochimica acta*, **37**, 651–66.

Kinzler, R. J. and T. L. Grove (1992) Primary magmas of mid-ocean ridge basalts:1. Experiments and methods. *Journal of Geophysical Research*, **97**, **B5**, 6885–906.

La Tourette, T., R. L. Hervig and J. Holloway (1995) Trace element partitioning between amphibole, phlogopite and basanite melt. *Earth and Planetary Science Letters*, **135**, 13–30.

Pfann, W. G. (1952) Principles of zone-melting. *Journal of Metals*, **4**, 747–53.

Shaw, D. M. (1979) Trace element melting models. In *Origin and Distribution of the Elements II*, ed. L. H. Ahrens. Oxford: Pergamon Press.

Vinogradov, A. P., A. A. Yaroshevsky and N. P. Ilyin (1971) A physico-chemical model of element separation in the differentiation of mantle material. *Philosophical Transactions of the Royal Society of London*, **A 268**, 409–21.

Yaroshevsky, A. A. (1970) Distribution of a component after one pass in zone melting with variable parameters. *Geochemistry International*, **7**, 979–81.

Zeck, H. P. (1970) An erupted migmatite from Cerro del Hoyazu. *Contributions to Mineralogy and Petrology*, **26**, 225–46.

9

Dynamic mantle melting

9.1 Introduction

Mantle melting takes place by one of three processes: (i) pressure release; (ii) temperature increase or (iii) addition of volatile components (McKenzie, 2000). This chapter is concerned largely with the formation of basaltic magmas, where pressure release during adiabatic mantle upwelling is the prime cause of melting. Trace-element behaviour will initially be examined in the case where some magma remains trapped within the melting mantle source, then later the mechanisms of melt migration will be considered.

9.2 Dynamic melting

The first topic is a melting process wherein some of the melt remains trapped within its source. This was considered in step (incremental) melting in the previous chapter, but now will be extended. Some definitions follow, additional to those already introduced in Chapter 7:

L, V^L	mass and volume of melt
W^s, V^W	mass and volume of solid
W^r, V^r	mass and volume of residue i.e. solid plus melt
f_p or φ	critical melt mass fraction when permeability is established
α	mass fraction melt/solid when permeability is established
ρ^s	density of rock matrix
ρ^l	density of liquid
ψ	volume porosity
β	mass proportion of material added in an open system, per unit of melting, F
γ	mass proportion of added material, β, contributed to the melt in an open system

Assuming that the uprising mantle begins to melt adiabatically, drops of initial melt form at grain contacts. As pointed out by Sobolev & Shimizu (1993, p. 183), the source 'cannot lose melt until the degree of melting attains a certain critical level'.

Maaloe (1982) had already described this critical level and referred to it as a *permeability threshold*, which depends on several factors, including the stress field and the viscosities of melt and source. Estimates of this critical melt fraction, here denoted by f_p, at which permeability is established, range from one part in a thousand to several %.

The source is assumed to be non-porous, because of the high pressure at the seat of melting; this is undoubtedly true for the mantle, which is the main focus here, but may be less true for crustal melting. Nevertheless, discussions of mantle melting frequently invoke porosity, e.g. Williams & Gill (1989, p. 1609) define *continuous melting* as 'continuous removal of magma from a progressively melting . . . source having a finite porosity. That is, we assume that some fraction of liquid produced during melting remains with the matrix and is not extracted'.

This appears to be a different concept from ordinary porosity, which relates to pore space which can be empty. Most writers about rock melting ignore this difference, and take the proportion of melt to define a volume porosity ψ, so that with definitions as given above

$$\psi = \frac{V^{\mathrm{L}}}{V^{\mathrm{L}} + V^{\mathrm{W}}} \tag{9.1}$$

When the mass fraction of liquid reaches a critical level, during melting, such that melt expulsion ensues, this level is sometimes called the *mass porosity*, φ or f_p, and it is readily shown that

$$\varphi = \frac{L^{\mathrm{p}}}{L^{\mathrm{p}} + W^{\mathrm{p}}} = f_p = \frac{\psi \rho^{\mathrm{l}}}{\psi \rho^{\mathrm{l}} + \rho^{\mathrm{s}}(1 - \psi)} \tag{9.2}$$

Some authors use φ for volume porosity, which can be confusing. Others prefer not to refer to porosity and, instead, use the ratio, α, of melt to solid, at the point where melt expulsion begins, where

$$\alpha = \frac{L^{\mathrm{p}}}{W^{\mathrm{p}}} \tag{9.3}$$

The difference between volume and mass fractions needs to be considered. The relationship of Eq. (9.2) is shown in Fig. 9.1, and it is seen that the difference is a small factor, even assuming considerable density difference between source and melt. The magnitude of the effect is exaggerated in this figure, because it is unlikely that a melt as light in density as 2.78 would come from a peridotite of density 3.5.

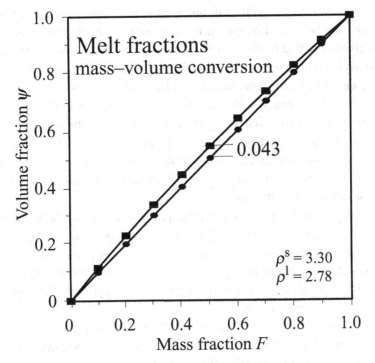

Fig. 9.1 When the source and the liquid to which it is melting have different densities, their volume and mass fractions differ. Their relationship is given by Eq. (9.2), shown here as a function of the chosen densities. The maximum difference of 0.043, or 8.6%, occurs at 50% melting, and corresponds to a density difference of about 16%.

The process under discussion was called *dynamic melting* by Langmuir *et al.* (1977), being based on the principle that 'melt is continuously removed from the mantle with some melt always remaining in the residue' (H. J. B. Dick).

Williams and Gill (1989, p. 1609) extended the definition to the situation where 'the system of matrix and interstitial fluid is moving instead of static. Fertile material (undepleted by melt extraction) constantly moves in to the source region' which may be contrasted with *continuous melting* as defined above. However the two terms will here be taken as indistinguishable and synonymous.

Equilibrium relations are usually assumed to govern the trace element concentration before the melt begins to migrate; the subsequent *dynamic (continuous) melting* with melt retained may be a non-equilibrium reaction.

The construction of a model therefore is made in terms of an equation for partial melting, such as Eq. (7.27)

$$\frac{dw^l}{w_0 - w^l} = \frac{1}{D} \cdot \frac{dF}{1 - F}$$

which is amenable to integration in terms of some measure of evolution of the system; this may be *time* or *temperature* or, as here, *F*, the degree of melting.

The degree of melting needs careful definition in circumstances where the source consists of an interstitial pre-existing melt in a solid matrix. The concept of 'melting' may apply either to (i) the solid material only, or to (ii) the whole source. In both cases, it is usual to assume that the melting solid and the new melt both equilibrate with the pre-existing melt, even if only momentarily. In the latter case it is necessary to modify the *D*-value (see below), to treat the pre-existing melt as another source phase (with partition coefficient of unity); this 'pre-existing melt' may in fact be solid material injected during some previous event e.g. basalt veins in peridotite, which have subsequently melted.

A fundamental choice when developing a model is whether the melting system is to be treated as open or closed. Since in every model some magma is being expelled, the system is open, but if material is not being added from elsewhere, then the system may conform in some ways to a closed model; this is the case for most of the discussion in previous chapters. However, explicitly open system models have also been proposed (see below).

Formulations for closed system melting have been made by several authors. McKenzie's model was put forward first (1984) in a fluid dynamics treatment of mantle melting, and subsequently (1985a,b) adapted specifically to trace element behaviour. His approach has been used by others (e.g. Williams and Gill, 1989; Eggins, 1992), who adapted it to the melting of a rising mantle plume. Although the solid was treated by McKenzie as monomineralic with a single constant *D*-value, he noted (1984, Eq. E30) that a polymineralic assemblage could be used, as is normally the case. In addition, no provision was made for variation of *D* with *F*, as occurs in non-modal melting.

In another formulation, Albarède (1995) shows that the modal melting of a partially solid source which already contains some previously formed melt can be modelled by treating the melt, m, as an extra phase with $K^{m-1} = 1$. This is analogous to the treatment of intercumulate melt during fractional crystallisation. The bulk partition coefficient for the solid phases, *D*, in the initial modal melting, is constant. The mass fraction of melt when permeability is established is f_p and so the bulk partition coefficient becomes

$$D^c = D(1 - f_p) + f_p K^{m-1} = D(1 - f_p) + f_p \qquad (9.4)$$

which is Albarède's Eq. (9.3.21). Fractional melting now follows the Rayleigh modal melting equation (Eq. 7.19), substituting D^c for *D*. Albarède's treatment may be converted to non-modal melting using Eq. (7.7).

Sobolev and Shimizu (1993) describe what they call a 'critical continuous melting (CCM)' model which is based on 'non-modal equilibrium melting . . . of mantle

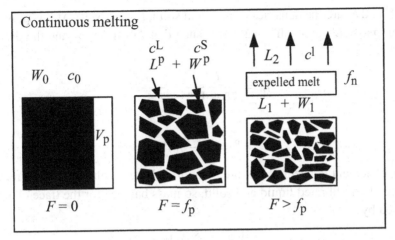

Fig. 9.2 Dynamic melting in a *'closed'* system. Initial batch melting proceeds until $F = f_p$, when there is a mass L^p of melt and W^p residual solid, at which point the source becomes permeable. Then *dynamic (continuous) melting* ensues; the mass ratio liquid/solid (α) remains constant and liquid L_2 is expelled, carrying the trace element concentration c^l. The fraction of *new* melt is $f_n = L_2/W_0 = L_2/(L^p + W^p) = F(1 + \alpha) - \alpha$, where F is the *total* melt fraction, including L_1.

material ... in a closed system (which) cannot lose melt until the degree of melting attains a certain critical level' (p. 183) where the restite (i.e. the residue) always retains the critical content of melt. The critical level has the same meaning as f_p, as already discussed.

Some of the difficulties encountered in the models just discussed have been examined by Zou (1998). After re-examination of modal melting along the lines used by McKenzie (1984, 1985a, 1985b), he adopts the Albarède approach to derive expressions for trace element variation during non-modal melting, and then for open system melting.

These models are reviewed in detail elsewhere (Shaw, 2000), and are in general agreement concerning trace element concentrations.

9.3 Closed system model

Before considering an open system model, it will be useful to lay out the principles of dynamic non-modal melting with trapped melt in a closed system, as just discussed.

Using the sketch in Fig. 9.2, a solid source of mass W_0 containing a trace element concentration c_0 begins batch melting, which continues to the critical level of volume permeability ψ. The mass permeability can be written as φ or f_p, and

$$f_p = \frac{\psi \delta^l}{\psi \delta^l + (1 - \psi)\delta^s} \tag{9.5}$$

where δ^l and δ^s are the densities of melt and solid, respectively. At this point there is a melt mass, L^p, and solid mass, W^p, such that $L^p/W^p = \alpha$, and the degree of melting is

$$f_p = \frac{L^p}{L^p + W^p} = \frac{\alpha}{1 + \alpha} \tag{9.6}$$

and

$$W^p = (1 - f_p)W_0 \tag{9.7}$$

The trace element concentrations are c^L and c^S. Batch melting may be treated as a mixing of equilibrated liquid and solid, so mass balance for the trace element is expressed by

$$Lc^L + Wc^S = W_0 c_0 \tag{9.8}$$

whence, using the bulk partition coefficient (Eq. 7.7), it follows that, for non-modal melting in the range $0 < F < f_p$

$$c^L = \frac{c_0}{D_0 + F(1 - P)} \tag{9.9}$$

The terminal values, when $F = f_p$, are c_p^L, c_p^S and D^p.

Dynamic or continuous melting now begins, and a further episode of melting f_2 takes place, leaving a solid residue W_1 where

$$W_1 = (1 - f_2)W^p \tag{9.10}$$

but, if the total melt fraction is F, then also

$$W_1 = (1 - F)W_0 \tag{9.11}$$

whence

$$f_2 = \frac{F - f_p}{1 - f_p} \tag{9.12}$$

A melt mass L_2 with concentration c^l is expelled (Fig. 9.2), leaving the residue $(L_1 + W_1)$, with mass ratio α, which henceforth remains constant.

The trace element mass balance is given by

$$w_0 = w^{L_2} + w^{L_1} + w^{W_1} = w^{L_2} + w^r \tag{9.13}$$

so, taking differentials

$$dw^{L_2} + dw^r = 0 \tag{9.14}$$

The bulk concentration in the source material residue now is c^r, and the concentrations in the melt and the solid are c^l and c^s, respectively, where c^l is the momentary concentration in the liquid. Consequently

$$c^l dL_2 + dw^r = 0 \tag{9.15}$$

But

$$w^r = c^l L_1 + c^s W_1 \tag{9.16}$$

and so

$$dw^r = L_1 dc^l + c^l dL_1 + W_1 dc^s + c^s dW_1 \tag{9.17}$$

whence

$$c^l dL_2 + L_1 dc^l + c^l dL_1 + W_1 dc^s + c^s dW_1 = 0 \tag{9.18}$$

The participating masses and their differentials are

$$L_2 + L_1 = F W_0 \tag{9.19}$$

$$W_1 = (1 - F)W_0 \qquad\qquad dW_1 = -W_0 dF \tag{9.20}$$

$$L_1 = \alpha W_1 = \alpha(1 - F)W_0 \qquad dL_1 = -\alpha W_0 dF \tag{9.21}$$

$$L_2 = f_n W_0 = (F(1 + \alpha) - \alpha)W_0 \qquad dL_2 = +(1 + \alpha)W_0 dF \tag{9.22}$$

$$W_r = L_1 + W_1 = (1 - F)(1 + \alpha)W_0 \quad dW_r = -(1 + \alpha)W_0 dF \tag{9.23}$$

and the equation becomes

$$c^l(1 + \alpha)dF + \alpha(1 - F)dc^l - \alpha c^l dF + (1 - F)dc^s - c^s dF = 0 \tag{9.24}$$

Also, since

$$c^s = c^l D \tag{9.25}$$

and

$$D = \frac{D_0 - PF}{1 - F} \tag{9.26}$$

so

$$dc^s = c^l dD + D dc^l \tag{9.27}$$

and

$$dD = \frac{D_0 - P}{(1 - F)^2} dF \tag{9.28}$$

and by substituting these in Eq. (9.24) we obtain

$$\frac{dc^l}{c^l} = \frac{(1 - P)dF}{D_0 + \alpha - F(P + \alpha)} \tag{9.29}$$

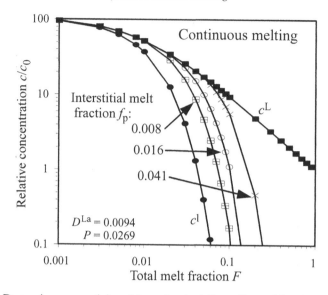

Fig. 9.3 Dynamic non-modal melting of spinel lherzolite, with mineral proportions, D_0 and P, as in Table 9.1. With three values chosen for the maximum
interstitial melt (Eq. 9.2) or permeability limit, the continuous melting curves for
La lie between those for batch melting (c^L) and fractional melting (c^l). One of the
curves has been extended beyond the point where melting would stop, because of
exhaustion of clinopyroxene ($F = 0.35$), to emphasise the trends.

Integration for the range $f_p < F < 1$, gives

$$\frac{c^l}{c_0^l} = \left(\frac{D_0 + \alpha - F(P + \alpha)}{D_0 + \alpha - f_p(P + \alpha)} \right)^{\frac{1-P}{P+\alpha}} \tag{9.30}$$

where

$$c_0^l = \frac{c_0}{D_0 + f_p(1 - P)} \tag{9.31}$$

Equation (9.30) is the same relationship as was derived by Sobolev and Shimizu
(1993), Albarède (1995) and Zou (1998). It is a three-parameter equation because
α is a function of f_p, as already seen. The variation of La concentration with total
melt fraction in a spinel lherzolite is seen in Fig. 9.3, calculated for three values
of the critical permeability limit, or interstitial melt, which correspond to volume
porosities of 0.01, 0.02 and 0.05. These continuous melting profiles lie between
equilibrium (c^L) and fractional (c^l) melting curves for permeable rock; that is,
if the source is permeable from the beginning of melting, then $f_p = \alpha = 0$ and
Eq. (9.30) becomes

$$c^l = \frac{c_0}{D_0} \left(1 - \frac{PF}{D_0} \right)^{\left(\frac{1}{P} - 1 \right)} \tag{9.32}$$

Table 9.1 *Bulk partition coefficients used in figures in Chapter 9*

	D_0	P
La	0.0094	0.0269
Ce	0.0152	0.0432
Pr	0.0244	0.0692
Nd	0.0337	0.095
Sm	0.0523	0.1474
Eu	0.06	0.1677
Gd	0.0675	0.1874
Tb	0.0755	0.2077
Dy	0.0832	0.2275
Ho	0.0813	0.2163
Er	0.081	0.2054
Tm	0.0912	0.2212
Yb	0.1035	0.2369
Lu	0.1094	0.2435
Ni	0.8924	6.58

These values were calculated using D^{i-m} in Table 7.1. and X_i^0 and p_i in Table 7.4

identical to Eq. (7.23), which is Rayleigh fractionation in a simple non-modal melting system.

The onset of permeability for a mantle peridotite probably corresponds to a small value of f_p, and so the concentration behaviour would be close to that for fractional melting; lesser permeability would lead to lesser fractionation.

A similar calculation for all the REE shows (Fig. 9.4) the difference in behaviour of the HREE, which are less incompatible in common peridotite minerals than the LREE. The HREE thus show much less divergence in the different melting mechanisms, as seen already in step melting theory (Chapter 7).

The concentration, c^l, of the trace element varies as the melt is expelled but, after the total degree of melting has reached F, the accumulated concentration is $\bar{c}$. The mass balance of the trace element may now be written

$$w^l = w_0 - w^s \tag{9.33}$$

so

$$\bar{c}(L_1 + L_2) = c_0 W_0 - c^s W_1 \tag{9.34}$$

whence

$$\bar{c} = \frac{1}{F}[c_0 - (1 - F)c^s] \tag{9.35}$$

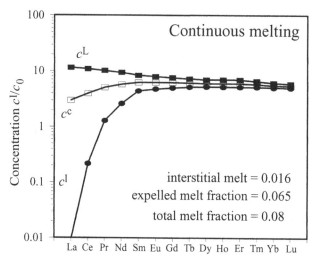

Fig. 9.4 Calculated relative concentrations of the REE for three types of non-modal melting of spinel lherzolite, for a mass porosity or permeability limit (f_p) of 0.016, and total melting degree $F = 0.08$, using D_0 and P values from Table 9.1. Dynamic melting gives concentrations (c^c) intermediate between batch and fractional melting, but the differences are small for the heavier, more compatible REE.

9.4 Open system model

As mentioned above, a number of authors have addressed melting in an open system, where source and melt may move independently at different velocities (Navon and Stolper, 1987; Iwamori, 1993; O'Hara, 1995a,b; Ozawa and Shimizu, 1995; Spiegelman, 1996; Vernières *et al.*, 1997 and Zou, 1998).

The simplest and essential feature of an open system in the present context is that solid material melts while matter is being added and melt is being released. A one-dimensional version is shown in Figure 9.5. The added mass at some degree of evolution (Fig. 9.5b) is denoted by W^{aa}, and the analysis is simplified if the rate of addition is assumed to be constant (see Ozawa and Shimizu, 1995) and equal to β per unit of melting F, so that

$$W^{aa} = \beta F^a W_0 \qquad (9.36)$$

It has not yet been specified whether the matter added is solid or liquid (Ozawa and Shimizu, 1995, refer to fluid). The choice is significant and at least three options are open. The first is addition of liquid or fluid which dissolves in available melt, but is not directly added to the solid phases. The second is addition of fluid, part of which dissolves in the melt, and part is added to the solids: this addition most likely takes place by metasomatism. The third is addition of both

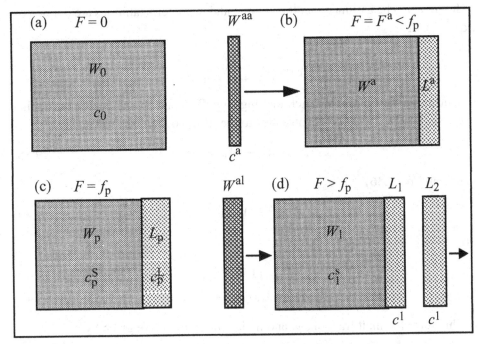

Fig. 9.5 Open system non-modal melting of an initial solid mass W_0, while a fluid mass W^{aa}, carrying a concentration c^a of the trace element, is being added to the system. Batch melting begins (a) in the solid system and then (b) at any degree of melting F^a less than f_p the melt mass is L^a. After the critical melt mass f_p has been reached (c) batch melting ends, melt begins to be expelled (L_2) with a concentration c^l (d), leaving an interstitial mass L_1 with the same concentration.

fluid and solid, both being incorporated into the melt, the latter by the process usually referred to as *assimilation*. Evidence exists for all three processes, but for the present we consider the first option. So it will be taken that the added material is *fluid*, that it contributes only to the melt,[1] and that there is momentary equilibration between this added matter, melt and solid. The added material is taken to be at the ambient temperature and pressure, and the enthalpy of mixing will not be considered.

The degree of melting, F, is defined in terms of the mass of melt produced from the original solid source, the remaining solid being

$$W = W_0(1 - F) \tag{9.37}$$

In this context, F is the degree of melt from the solid source, but it is NOT the total melt fraction, for fluid is being added and melt is being expelled.

---

[1] That is, the factor γ (see p. 211) is equal to unity.

It will again be assumed that the volume and mass permeabilities are defined by Eq. (9.5), or

$$f_p = \frac{\psi \delta^l}{\psi \delta^l + (1 - \psi)\delta^s}$$

During the first phase of batch melting (Fig. 9.5a and b), the degree of melting is F^a, and the matter added contributes to the melt, so the mass balance is

$$L^a + W^a = W_0 + W^{aa} \tag{9.38}$$

and from Eq. (9.36)

$$L^a + W^a = W_0(1 + \beta F^a) \tag{9.39}$$

So

$$W^a = (1 - F^a)W_0 \tag{9.40}$$
$$L^a = W_0 F^a(1 + \beta) \tag{9.41}$$

At this stage the melt fraction, φ, and melt/solid ratio, α, are given by

$$\varphi = \frac{\text{melt}}{\text{total}} = \frac{F^a(1 + \beta)}{1 + \beta F^a} \tag{9.42}$$

$$\alpha = \frac{\text{melt}}{\text{solid}} = \frac{F^a(1 + \beta)}{1 - F^a} \tag{9.43}$$

The mass balance for mineral i may be written (see Eq. 7.4) as

$$W^a X_i = W_0 X_i^0 - p_i L^a$$

so, using Eq. (7.7)

$$D = \sum X_i D^{i-1} = \frac{D_0 - F^a P(1 + \beta)}{1 - F^a} \tag{9.44}$$

The partition coefficient is thus affected by the material added, and becomes a variable even for modal melting; for, if $P = D_0$ then

$$D = D_0 \cdot \frac{1 - F^a(1 + \beta)}{1 - F^a} \tag{9.45}$$

This is illustrated in Fig. 9.6, which shows how dependent D is on β. For example, at 10% melting the effective partition coefficient has decreased by about 10% if the rate of fluid addition equals the degree of melting (i.e. $\beta = 1.0$). The curve for $\beta = 10$ is included to show this dependence, even though it is difficult to envisage an influx of this magnitude (which would more closely resemble the behaviour of a fumarole).

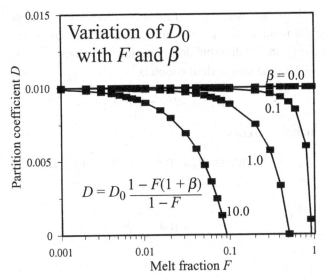

Fig. 9.6 During open-system melting the partition coefficient of a trace element is affected by the material added, which changes the mineral proportions. Even the value D_0 for modal melting is variable and depends on the proportionality factor β. The change in every case *decreases* the coefficient.

Note that D is variable *whether or not the material added carries the trace element*. This is because the material influx affects the mass balance of each mineral in the system, and so the value of D_0 is controlled mainly by the parameter β. It is also evident that D *must* decrease, and so increase the degree of incompatibility of incompatible trace elements or, for an element such as Ni, decrease the degree of compatibility.

Taking f_p again as the batch melting limit before melt is expelled, then for the range $0 < F < f_p$, while equilibrium melting is taking place, the trace element behaviour is governed by

$$c^L L^a + c^S W^a = c_0 W_0 + c^a W^{aa} \tag{9.46}$$

so

$$c^L = \frac{c_0 W_0 + c^a W^{aa}}{L^a + D W^a} = \frac{c_0 + c^a \beta F}{D_0 + F(1 + \beta)(1 - P)} \tag{9.47}$$

where c^a is the trace element concentration in the material being added. At the limit, when $F = f_p$, the melt fraction φ and melt ratio α are given by modifying Eq. (9.43)

$$\varphi = \frac{f_p(1 + \beta)}{1 + f_p \beta} \qquad \alpha = \frac{f_p(1 + \beta)}{1 - f_p} \tag{9.48}$$

The concentrations c_p^L and c_p^S are given by Eq. (9.48) and by $D c_p^L$ respectively.

Continued melting (i.e. when $F > f_p$) leads to expulsion of melt L_2 (Fig. 9.5c and d) leaving the residue, W^r, consisting of L_1 and W_1, whose ratio, α, remains constant as melting and melt expulsion continue. The total melt mass comprises melted source plus the matter added, which is

$$W^{aa} + W^{al} = \beta F W_0 \tag{9.49}$$

So, the mass balance becomes

$$L_1 + L_2 + W_1 = W_0 + W^{aa} + W^{al} = W_0(1 + \beta F) \tag{9.50}$$

But, since

$$W_1 = W_0(1 - F) \tag{9.51}$$

consequently

$$L_1 + L_2 = W_0 F(1 + \beta) \tag{9.52}$$

and

$$L_1 = \alpha W_0(1 - F) \tag{9.53}$$
$$L_2 = W_0(-\alpha + F(1 + \alpha + \beta)) \tag{9.54}$$
$$W^r = L_1 + W_1 = W_0(1 + \alpha)(1 - F)$$

Also

$$D = \frac{D_0 - FP(1 + \beta)}{1 - F} \tag{9.55}$$

If the additional degree of melting (beyond f_p) is f_2 then it is easy to show that, as in Eq. (9.11) in the closed system

$$f_2 = \frac{F - f_p}{1 - f_p} \tag{9.56}$$

Before continuing, it is useful to survey the relative magnitudes of the masses and their ratios in an open melting system, dependent as they are on the parameters f_p and β, and on the variable F. In Fig. 9.7 the influx rate $\beta = 0.10$, and when complete melting has occurred, that is when $F = 1.0$, the influx mass $W^{aa} + W^{al}$ is now 0.10, and the total mass of the system has thus increased to 1.10. Also, since all of the fluid influx goes into the liquid, the final mass of expelled liquid, L_2, reaches 1.10, and the residual solid mass, W_1 is 0.00. Here the idea of complete melting describes the fate of the original mass of source rock, and this will be reflected in the trace element behaviour.

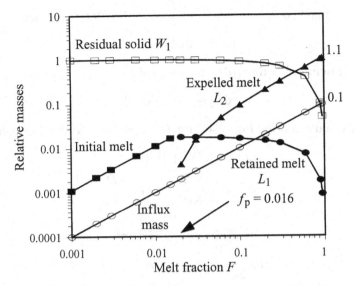

Fig. 9.7 The relative masses of different parts of an open melting system depend on the parameters β and f_p, and the variable F. The melting concludes when $F = 1.0$, and in the case shown here, where β has been chosen as 0.10, the melt mass is now 1.10 because of the addition of the fluid.

Returning to the mass balance, the differentials of the various masses are

$$dW^{al} = \beta W_0 dF \tag{9.57}$$

$$dW_1 = -W_0 dF \tag{9.58}$$

$$dL_1 = -\alpha W_0 dF \tag{9.59}$$

$$dL_2 = (1 + \alpha + \beta) W_0 dF \tag{9.60}$$

$$dW^r = -(1 + \alpha) W_0 dF \tag{9.61}$$

The trace element mass balance is, from Eq. (9.50),

$$w^{L_1} + w^{L_2} + w^{W_1} = w^{L_2} + w^r = w_0 + w^{aa} + w^{al} \tag{9.62}$$

where

$$w^{L_2} = \int_{f_p}^{F} c^l dL \quad w^r = c^r W^r \quad w^{aa} + w^{al} = c^a \beta F W_0 \tag{9.63}$$

c^a is the concentration in the influx mass, and c^r is the concentration in the residue, so

$$c^r = \frac{c^l L_1 + c^s W_1}{L_1 + W_1} = c^l \cdot \frac{\alpha + D}{1 + \alpha} \tag{9.64}$$

To express the mass conservation we differentiate Eq. (9.62) to obtain

$$dw^{L_2} + dw^r = dw^{al} \tag{9.65}$$

and, since we define $c^l = dw^{L_2}/dL_2$ then we have

$$c^l dL_2 + dw^r = dw^{al} \tag{9.66}$$

This can be evaluated using Eq. (9.44) for D and the following relationships

$$dw^r = c^r dW^r + W^r dc^r \tag{9.67}$$

$$dc^r = \frac{\alpha + D}{1 + \alpha} dc^l + c^l \frac{dD}{1 + \alpha} \tag{9.68}$$

$$dD = \frac{D_0 - P(1 + \beta)}{(1 - F)^2} dF \tag{9.69}$$

$$dw^{al} = c^a \beta W_0 dF \tag{9.70}$$

With substitution of these differentials, Eq. (9.66) becomes

$$\frac{dc^l}{c^l(1 + \beta)(1 - P) - c^a \beta} = -\frac{dF}{D_0 + \alpha - F[\alpha + P(1 + \beta)]} \tag{9.71}$$

and on integration over the range f_p to F we obtain

$$\frac{c^l}{c_0} = \left(\frac{c_0^l}{c_0} - y \cdot \frac{c^a}{c_0} \right) \cdot \left(\frac{1 - zF}{1 - zf_p} \right)^u + y \cdot \frac{c^a}{c_0} \tag{9.72}$$

where,[2]

$$y = \frac{\beta}{(1 + \beta)(1 - P)} \tag{9.73}$$

$$u = \frac{(1 + \beta)(1 - P)}{\alpha + P(1 + \beta)} \tag{9.74}$$

$$z = \frac{\alpha + P(1 + \beta)}{D_0 + \alpha} \tag{9.75}$$

Also, $c_0^l = c_p^L$, as given in Eq. (9.47) and $c^s = Dc^l$. It may be noted that if the proportion of the added matter is vanishingly small (so that $\beta = 0$), then Eq. (9.72) collapses into Eq. (9.30).

The accumulated concentration, $\bar{c}$, in all the melt expelled, is obtained as before by modifying Eq. (9.62) as follows

$$w^l = \bar{c}(L_1 + L_2) = w_0 + w^{aa} + w^{al} - w^s \tag{9.76}$$

---

[2] There is an error in similar equations as given in Shaw (2000, p. 1058).

thus obtaining

$$\bar{c} = \frac{c_0 + c^a \beta F - c^s(1 - F)}{F(1 + \beta)} \tag{9.77}$$

9.5 Discussion of models

The behaviour of Ce, a strongly incompatible element, in open-system melting of peridotite, is shown in Fig. 9.8a for a particular choice of the parameters in Eq. (9.72). The melt concentration is similar to that in a closed system (see Fig. 9.3) when F is less than or only slightly greater than f_p, but then changes course, depending on the values held by c^a and β. By contrast, an incompatible element such as Ni shows little difference in behaviour from the closed-system trends, even when, as in Fig. 9.8b, the influx concentration is as great as ten times that in the melting rock. The effect of variation in c^a, when β is held constant, is shown in Fig. 9.9; when the added matter is much richer in an incompatible element (Fig. 9.9a) the concentration in the melt is buffered, but this is not the case for a compatible element (Fig. 9.9b).

Increasing the rate of addition, β, reinforces the effect of either high or low c^a, not only for a highly incompatible element like Ce (Fig. 9.10), but also for more compatible ones such as Yb or Ni, so the choice of each of these parameters has a powerful effect on open-system modelling. With the more compatible element the effect of variations in β are minor, except at very large c^a values, which may be unrealistic.

It may also be noted from Figs. 9.8 to 9.10 that the behaviour of a trace element is more influenced by the influx concentration, c^a, and the rate of influx β, than by the coefficients D_0 and P.

When considering groups of related elements such as the REE, the comparison of different degrees of melting, for a chosen set of parameters (Fig. 9.11a), shows the relative fractionations which can occur as a function of compatibility differences. In this figure the value of c^a is the same for each element, to emphasise the fractionation effects, but this is unrealistic. By contrast, in a more realistic situation (Fig. 9.11b), the upper points have been calculated using an infiltrating fluid enriched in the LREE relative to the HREE; for each element the value used for c^a was an alkali basalt concentration used by Bodinier *et al.* (1990) as a source of metasomatism and melt infiltration reacting on solid peridotite. Even with an influx factor of only 10%, these lead to an order of magnitude increase for the LREE (note the logarithmic scale).

In the previous section it was assumed that fluid material was added solely to the melt. In the case where the fluid is also added to the solid it is necessary to

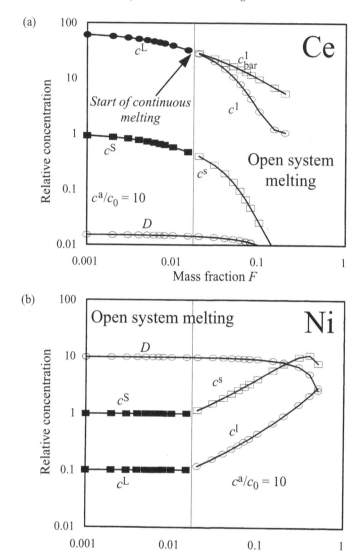

Fig. 9.8 (a) The concentration of Ce during non-modal open-system melting of spinel lherzolite with D_0 and P, as in Table 9.1, for a given rate of fluid addition ($\beta = 0.1$) and trace element addition ($c^a/c_0 = 10$); (b) behaviour of the incompatible element Ni.

specify the mass proportion $(1 - \gamma)$; the proportion added to the melt is γ. Then, following from Eq. (9.36), we divide the matter added into two portions, liquid and solid. Thus we can write

$$W^{aa} = \text{liquid} + \text{solid} = L^{aal} + W^{aal} \tag{9.78}$$

$$L^{aal} = \gamma W^{aa} \quad W^{aal} = (1 - \gamma)W^{aa} \tag{9.79}$$

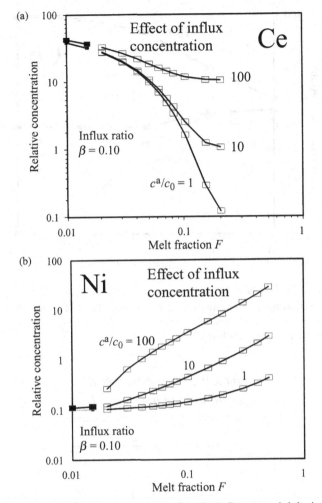

Fig. 9.9 The effect of a high concentration c^a in the influx material during the open system melting shown in Fig. 9.8 is to buffer the melt concentration and suppress fractionation for an incompatible element (a). For a compatible element (b), a high influx concentration increases fractionation.

If the total mass of melt is L^a, then

$$L^a = F W_0 + \gamma W^{aa} = F(1 + \gamma\beta)W_0 \tag{9.80}$$

Similarly

$$W^a = (1 - F(1 - \beta + \gamma\beta)) W_0 \tag{9.81}$$

so that

$$L^a + W^a = (1 + \beta F)W_0 \tag{9.82}$$

as before.

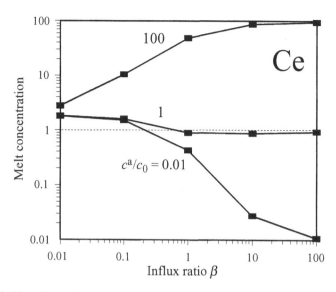

Fig. 9.10 The effect of both c^a and β on the melt concentration of an incompatible trace element during 10% melting of the same system as in Fig. 9.8. Large values of β, of course, would correspond to a process similar to magma mixing, and are not appropriate to mantle melting.

Table 9.2 *Normalised REE*
concentrations in infiltrating fluid

	c^a/c_0
La	190
Ce	150
Nd	90
Sm	50
Eu	38
Tb	20
Yb	10
Lu	9

from Bodinier *et al.*, 1990, Fig. 10.

Further development along these lines will not be continued here. The 'solid' component added is really the product of a metasomatic reaction, dominated by element diffusion, and must be so treated (see for example Bodinier *et al.*, 1990).

The foregoing paragraphs present a statement of the formalism needed for modelling the melting of dynamic closed- and open-mantle systems, and the consequences of the choice of parameters. The same approach can be readily applied to

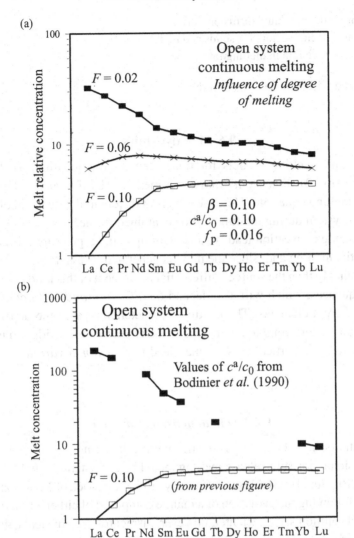

Fig. 9.11 (a) Three different degrees of open system melting, as in Fig. 9.8, with constant influx factor (β) and constant influx concentration (c^a) for each of the REE. In (b) parameters are the same as in (a), except the additional influx concentrations are the alkali basalt values given in Table 9.2, with greatly increased LREE concentrations.

crustal melting (but see Bea, 1996). Some factors which have *not* been included are:

(i) partition coefficient variability;
(ii) non-cotectic mineral proportions;
(iii) mineral reactions (i.e. incongruent melting);

(iv) differing solid and liquid diffusion times;
(v) differing velocity vectors for solids and melts;
(vi) directional variation in melting rates.

The last three will now be considered.

9.6 Melt dynamics

In the dynamic melting models for trace element fractionation, as defined in
Eqs. (9.36) to (9.72), the system behaviour is measured by F, which is the only variable in the melting equations. Here the effects of differential movement of different
parts of the system during melting will be examined further.

When mantle convection leads to a zone of upwelling, pressure release will in-
stigate partial melting and, since the magma generated is less dense, it will tend to
rise more quickly than the source. Differential movement of this kind has geochemical consequences, which will also depend on *differing solid and liquid diffusion
rates* for different elements. The melting parameters now are time-dependent and
the evolution is more complex. Differing velocity vectors for solids and melts also
are not limited to a vertical axis, so may lead to *directional variation in melting
rates*.

9.6.1 One-dimensional motion

Several authors have considered mantle melting in terms of the evolution of a
one-dimensional pressure-release melting model in a gravitational field. Both Ribe
(1985) and Richter (1986) follow the fluid dynamics approach of McKenzie (1984),
the former following the evolution of a mantle composed of either a binary eutectic
or of a solid solution series. Richter, however, disregards the physical nature of the
melting matrix.

The simplest system of this kind can be visualised as the one-dimensional upwelling zone shown in Fig. 9.12, which is a more developed version of Fig. 9.2.
Richter has analysed the ways in which chemical fractionation can take place in
this and more complex systems. His approach assumes modal melting, constant
partition coefficients and local diffusive equilibrium, and develops equations for
perfect equilibrium and for perfect fractional partial melts. The behaviour of the
system is expressed in several ways, all of which contrast the upward velocity w
of the buoyant melt with the downward velocity W of the subsiding residual solid
source.

The behaviour of a trace element is best expressed as the evolution of its liquid
concentration c^l as melting proceeds. This is controlled by variables additional to

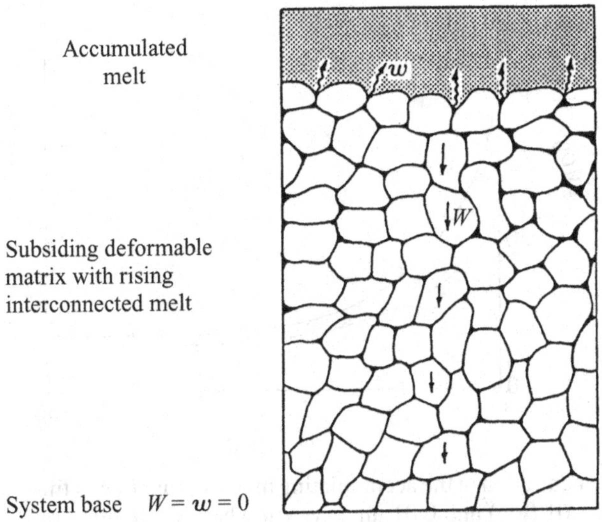

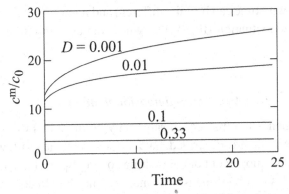

Accumulated
melt

Subsiding deformable
matrix with rising
interconnected melt

System base $W = w = 0$

Fig. 9.12 A granular, deformable source undergoes melting. The buoyant melt is interconnected for any value of the melt fraction and rises with velocity w as the source subsides with velocity W, towards the base, where $W = 0$. The expelled melt accumulates in a reservoir or is erupted directly (after Richter, 1986, Fig. 1).

Fig. 9.13 The concentration in the accumulating melt of four elements of different D-values in a system where melting is faster than the rate of melt segregation; the concentration rises with time, particularly for the most incompatible elements (after Richter, 1986, Fig. 4a).

the fraction of liquid F, in particular by the matrix subsidence rate (W) relative to the melt rising rate (w). When melting is rapid, relative to the rate of segregation, c^l in the expelled or accumulated melt *increases with time* (Fig. 9.13) for incompatible elements. This variation of concentration in the melt is seen in Fig. 9.13 for four elements of different D-values; in this case the concentrations of the incompatible elements increase with time.

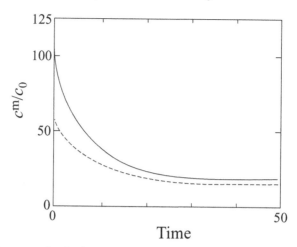

Fig. 9.14 Concentration in the accumulating melt as a function of time, for partition coefficients 0.001 (top) and 0.01, in a system where the melting rate is similar to the segregation rate. The concentration for the most incompatible element (top curve) decreases with time (after Richter, 1986, Fig. 12b).

By contrast, in a system where segregation occurs as fast as melting, the evolution of concentration with time can be quite different and it may, for instance, be possible to find the reversed situation (Fig. 9.14), where the concentrations diminish with time.

9.6.2 Two-dimensional motion

Mantle melting beneath a mid-ocean ridge may be modelled using the buoyancy upwelling approach already described, but it is more realistic to allow for a horizontal component to the movement of the melting source. Spiegelman (1996) analyses the flow mechanics of two-dimensional melting and the effects on trace-element behaviour. The melt and solid flow fields were calculated by Spiegelman for two dynamic cases, which he calls *active* and *passive*. The former corresponds to plate movement whereas the latter is an extension of the buoyancy upwelling or one-dimensional case.

The relations between melting rate and solid flow trajectory are functions of depth and distance from the ridge and are shown, for appropriate choice of melting parameters, in Fig. 9.15; the melting rates decrease away from the ridge. Section (a) is the active case and (b) is the passive one. In case (a) the melting rate contours show how melting is driven upwards towards the ridge zone, and Spiegelman finds that the upwelling velocity is many times greater than in the passive case.

(a)

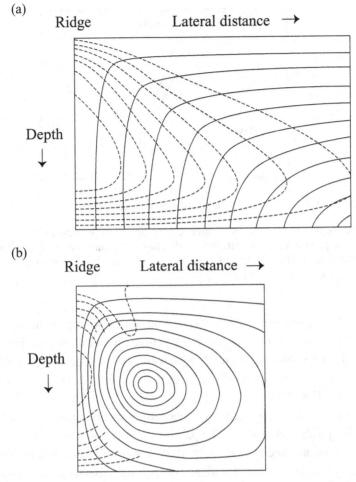

Fig. 9.15 Variation of solid flow (solid curves) and melt rate (dashed curves) for active (a) and passive (b) mantle melting beneath a mid-ocean ridge (after Spiegelman, 1996, Fig. 1). The two figures are vertical cross-sections showing depth and surface distance from the ridge (upper left).

Trace-element behaviour is considered for *equilibrium* transport where melt and solid are in equilibrium at all times, and *disequilibrium* transport, where the solid melts instantaneously to an equilibrium melt, but does not interact with any of the melt percolating through it. In equilibrium transport a completely incompatible element ($D = 0$) will travel along the melt flow lines in Fig. 9.15 whereas a completely compatible element ($D = \infty$) will travel along the solid flow lines. Spiegelman derives the equations of trace-element transport for the active and passive cases, for equilibrium and disequilibrium melting, and Fig. 9.16 shows how melt concentrations (c^l/c_0) of different elements (normalised to the bulk source)

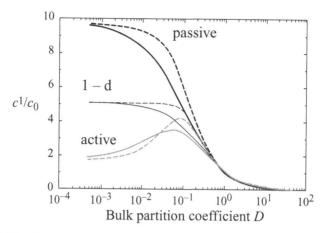

Fig. 9.16 Liquid concentration in melt emerging along a ridge axis as a function of bulk partition coefficient, for equilibrium (full curve) and disequilibrium
(dashed curve) conditions, during passive, active and one-dimensional melting
(after Spiegelman, 1996, Fig. 3).

vary with the bulk partition coefficient (D), assuming a homogeneous source; continuous curves are for equilibrium and dashed curves are for disequilibrium melting.
The concentrations shown are for melts which erupt or emerge along the ridge axis.
The one-dimensional curves are for a steady-state vertical melting column with
melting fraction $F = 20\%$.

For very incompatible elements the concentrations are similar for either equilibrium or disequilibrium models. The passive melting curves correspond to lower
degrees of melting, and accordingly show that the most incompatible elements have
much higher concentrations than with active melting behaviour. A curious feature
of active melting, however, is to find peak concentrations at intermediate degrees of
incompatibility, not at extreme incompatibility. As Spiegelman (1996, p. 121) says,
'this behavior is not found in standard geochemical models', referring of course to
most of the models discussed earlier in this book.

The analysis continues with an examination of concentration variation in melts
erupted or intruded away from the ridge axis, and then to a comparison of transport
with non-transport models, to emphasise 'the geochemical consequences of fluid
mechanics' (p. 126), concluding with a discussion of applications to the topic of
MORB genesis.

An evaluation of mantle melting in a situation similar to Fig. 9.15a has also been
made by O'Hara (1985). The maximum melt production is at the ridge or central
axis, and falls off along the melt lines ('stream' lines in O'Hara). Melt produced at
different points along the melt lines accumulates but it is assumed that it is erupted

or extruded only along the central axis. O'Hara calls the geometry of this cross-section pattern the 'shape' of the melting regime, and he presents ways in which the trace element concentrations may be estimated.[3] O'Hara assumes that the parcel of mantle melting to a degree F_y, at a (horizontal) distance y along a stream line is simply related to the maximum F_m, at distance Y_m to the ridge axis by

$$y/Y_m = (F_y/F_m)^n \qquad (9.83)$$

without defining n; the melt fractions F_y and F_m, are taken as proportional to h and h_m. Then, assuming a constant bulk partition coefficient for the solid mantle, the trace element concentration formed by fractional melt accumulation,[4] is given by the familiar Eq. (7.24). The parcel of melt thus formed undergoes no further reaction, but is mixed with all the others and becomes available for eruption at the ridge axis. O'Hara presents a procedure for calculating the mix but the details, which depend on the geometry of the flow lines, are not articulated.

Three-dimensional melting is examined by Eggins (1992) and is found necessary to explain REE patterns in Hawaiian basalts.

9.6.3 Percolation of melt through mantle

The previous paragraphs have outlined melting progress in mobile mantle, the resultant magmas erupting or accumulating at a ridge axis or centre. It may be that such magma continues to rise and percolate through overlying mantle whose temperature is below the liquidus. Interaction of melt with a cooler mantle can lead to mineralogic changes, such as those documented by Kelemen *et al.* (1992) for the Trinity peridotite in northern California.

They describe older pyroxenite bands which have been transformed to olivine-rich dunite and harzburgite, which carry chromian spinel, just as do the pyroxenites, whose continuity is not broken. The dunite and harzburgite are *replacive* in reaction zones which developed without volume change, on a scale of centimetres to metres, as they have been described in other ophiolites. The authors explain that *in situ* melting cannot be entertained as an explanation and infer that such reaction zones 'form by interaction between solid peridotite and silicate melt infiltrating along grain boundaries' (p. 635). The authors document this process in great detail with major and trace element evidence, and believe that it can occur on a much wider

---

[3] He gives no formal definition of 'shape factor', but indicates on p. 3 that it is a function of the power n alone.

[4] O'Hara also assumes that some melt remains trapped in the mantle residue.

scale to explain that 'mantle harzburgites and opx-rich peridotites cannot be the residua of partial melting from a primitive mantle' (p. 640).

This percolation and reaction process is further explored in two later publications (Kelemen *et al.*, 1995, 1997) directed largely to the origin of mid-ocean ridge basalts. The authors present persuasive evidence that dunites in ophiolites are in close equilibrium with MORB and have formed by melt percolation (which they term 'focussed flow'), and that the lherzolites are an intermediate facies in the formation of the dunites.

A different approach to melt percolation reactions, still dependent on the different concentrations of an element in the liquid and in the solid, has been modelled on the principle of the ion-exchange column. Navon and Stolper (1987) argue that the concentration (c_i^l) of each trace element (i), as the melt column moves upwards, depends on its partition coefficient (D_i^{s-l}) and the melt velocity (v^l) relative to the solid.

Assuming that both local equilibrium, as given by $D_i^{s-l} = c^s/c^l$, and diffusion in liquid and solid are instantaneous, then in a one-dimensional column of uniform initial composition, the authors find that the concentration c_i^l changes with time t and distance z along the column by the relation

$$\frac{dc_i^l}{dt} + X_i^l v^l \frac{dc_i^l}{dz} = 0 \tag{9.84}$$

where X_i^l is the proportion of the trace element in the melt at equilibrium

$$X_i^l = \frac{a}{a + b D_i^{s-l}} \tag{9.85}$$

where a and b are constants. This may be re-arranged to an equation which describes the rate at which levels of constant concentration at z moves through the melt column, as follows

$$\frac{dz}{dt} = X_i^l v^l \tag{9.86}$$

Since $X_i^l < 1$, it follows that trace elements are transported through the column more slowly than the melt is moving. Also, since X_i is a function of the inverse of D_i^{s-l}, the compatible elements (high D_i^{s-l}) move more slowly than incompatible ones.

It is inferred that levels of constant concentration (fronts) for different trace elements will be situated at different distances from the origin (the base of the column). When the melt has travelled a distance Z, the front for element i will be at a distance

$$z_i = X_i^1 Z$$

Fronts for different elements will reach the top of the column at different times and consequently a sequence of erupted magma flows will vary in trace element concentrations. This conclusion is difficult to accept, perhaps because the operating process is not identical with an ion-exchange column, and consequently Eq. (9.84) cannot be applied to a situation where reaction between solid and fluid is taken as instantaneous.

9.7 Summary

During mantle melting both solid and liquid material are moving and, since the magma generated is less dense, it will tend to rise more quickly than the source. Differential movement of this kind has geochemical consequences. Early melt remains trapped within the source, and cannot be expelled until the degree of melting attains a certain critical level referred to as a *permeability threshold*, which depends on several factors, including the stress field and the viscosities of melt and source. This critical melt fraction at which permeability is established, in such *dynamic* or *continuous* melting, is sometimes called the *mass porosity*, φ or f_p, and is one of the parameters in the melting equation.[5]

The behaviour of an incompatible element during dynamic non-modal melting is intermediate between batch melting and fractional melting. Melting starts on the batch curve, then departs from it as permeability is exceeded; the smaller the mass porosity, the closer to fractional melting behaviour. The magnitude of the porosity also affects the concentration in the melt substantially, e.g. doubling the porosity from 1.5 to 3% can lead to an order of magnitude concentration difference when about 10% melting has taken place. For compatible elements these effects are less pronounced, as may be seen in the behaviour of the LREE contrasted with the HREE.

In the process of *open-system dynamic melting* addition of material takes place. Here it is assumed that the material enters in a fluid phase which is incorporated into the melt, while reacting (equilibrating) with coexisting solids. In addition to f_p, two more parameters are needed to model trace element concentration: c^a is the trace element concentration in the fluid phase, and β is the rate at which that phase is being added to the system. When permeability is exceeded the trace-element concentration begins to diverge from batch behaviour. The degree of divergence

---

[5] The others being D_0 and P, as in the previous chapter.

depends on the magnitudes of c^a and β, and is more marked for incompatible than compatible elements.

In modelling the behaviour of a group of trace elements, such as the rare earths, it is of course necessary to know, or make assumptions about, the concentration of each in the primary material melting (c_0) as well as in the fluid phase (c^a).

When source and melt are both in motion they may react. The simplest case will be a one-dimensional (vertical) system, where new melt encounters unmelted source material. Trace element concentrations in the melt will increase or decrease with time, depending in part on its relative velocity. Two-dimensional flow can be more complex and incompatible elements may follow different paths from compatible ones. Reactions during percolation of melt through mantle are integral stages of basalt genesis.

References

Albarède, F. (1995) *Introduction to Geochemical Modeling.* Cambridge: Cambridge University Press, 543pp.

Bea, F. (1996) Controls on the trace element chemistry of crustal melts. *Geological Society of America special paper*, **315**, 33–41.

Bodinier, J. L., G. Vasseur, J. Vernières, C. Dupuy and J. Fabries (1990) Mechanisms of mantle metasomatism: geochemical evidence from the Lherz orogenic peridotite. *Journal of Petrology*, **31**, 597–628.

Eggins, S. M. (1992) Petrogenesis of Hawaiian tholeiites: 2, aspects of dynamic melt segregation. *Contributions to Mineralogy Petrology*, **110**, 398–410.

Iwamori, H. (1993) Dynamic disequilibrium melting model with porous flow and diffusion-controlled chemical equilibration. *Earth and Planetary Science Letters*, **114**, 301–13.

Kelemen, P. B., H. J. B. Dick and J. E. Quick (1992) Formation of harzburgite by pervasive melt/rock reaction in the upper mantle. *Nature*, **358**, 635–41.

Kelemen, P. B., N. Shimizu and V. J. M. Salters (1995) Extraction of mid-ocean-ridge basalt from the upwelling mantle by focused flow of melt in dunite channels. *Nature*, **375**, 747–54.

Kelemen, P. B., G. Hirth, N. Shimizu, M. Spiegelman and H. J. B. Dick (1997) A review of melt migration processes in the adiabatically upwelling mantle beneath oceanic spreading ridges. *Philosophical Transactions of the Royal Society London*, **Ser. A, 355**, 283–318.

Langmuir, C. H., A. F. Bender, A. E. Bence, G. N. Hanson and S. R. Taylor (1977) Petrogenesis of basalts from the FAMOUS area, Mid-Atlantic Ridge. *Earth and Planetary Science Letters*, **36**, 133–56.

Maaloe, S. (1982) Geochemical aspects of permeability controlled partial melting and fractional crystallization. *Geochimica et cosmochimica acta*, **46**, 43–57.

Maaloe, S. and A. D. Johnston (1986) Geochemical aspects of some accumulation models for primary magmas. *Contributions to Minerology and Petrology*, **93**, 449–58.

McKenzie, D. (1984) The generation and compaction of partially molten rock. *Journal of Petrology*, **25**, 713–65.

(1985a) $^{230}Th - ^{238}U$ disequilibrium and the melting processes beneath ridge axes. *Earth and Planetary Science Letters*, **72**, 149–57.

(1985b) The extraction of magma from the crust and the mantle. *Earth and Planetary Science Letters*, **74**, 81–91.

(2000) Constraints on melt generation and transport from U-series activity ratios. *Chemical Geology*, **162**, 81–94.

McKenzie, D. and R. K. O'Nions (1991) Partial melt distributions from inversion of REE concentrations. *Journal of Petrology*, **32**, 1021–91.

Navon O. and E. M. Stolper (1987) Geochemical consequences of melt percolation, the upper mantle as a chromatographic column. *Journal of Geology*, **95**, 285–307.

O'Hara, M. J. (1985) Importance of the 'shape' of the melting regime during partial melting of the mantle. *Nature*, **314**, 58–62.

(1995a) Imperfect melt separation, finite increment size and source region flow during fractional melting and the generation of reversed or subdued discrimination of incompatible trace elements. *Chemical Geology*, **121**, 27–50.

(1995b) Trace element geochemical effects of integrated melt extraction and 'shaped' melting regimes. *Journal of Petrology*, **36**, 1111–32.

Ozawa, K. and N. Shimizu (1995) Open-system melting in the upper mantle: constraints from the Hayachine-Miyamori ophiolite, northeastern Japan. *Journal of Geophysical Research*, **100, B11**, 22315–35.

Ribe, N. (1985) The generation and composition of partial melts in the Earth's mantle. *Earth and Planetary Science Letters*, **73**, 361–76.

Richter, F. M. (1986) Simple models for trace element fractionation during melt segregation. *Earth and Planetary Science Letters*, **77**, 333–44.

Shaw, D. M. (2000) Continuous (dynamic) melting theory revisited. *Canadian Mineralogist*, **38, 5**, 1041–63.

Sobolev, A. V. and N. Shimizu (1993) Superdepleted melts and ocean mantle permeability. *Transactions of the Russian Academy of Sciences. Earth Science Sections*, **328.1**, 182–8 (translated from (1992), *Dokl. ross. akad. nauk, SSSR, Ser. A*, **326.2**, 354–60).

Spiegelman, M. (1996) Geochemical consequences of melt transport in 2-D; the sensitivity of trace elemenets to mantle dynamics. *Earth and Planetary Science Letters*, **139**, 115–32.

Vernières, J., M. Godard and J.-L. Bodinier (1997) A plate model for the simulation of trace element fractionation during partial melting and magma transport in the Earth's upper mantle. *Journal of Geophysical Research*, **102**, 24771–2.

Williams, R. W. and J. B. Gill (1989) Effects of partial melting on the uranium decay series. *Geochimica et cosmochimica acta*, **53**, 1607–19.

Zou, H. (1998) Trace element fractionation during modal and non-modal dynamic melting and open-system melting; a mathematical treatment. *Geochimica et cosmochimica acta*, **62**, 1937–46.

Subject index

accessory minerals, 181
accuracy, 3
AFC, 105, 121
allanite, 23, 24
analytical methods, 10
assimilation, 94, 99, 103, 134, 221
assimilation-fractional crystallisation (AFC) process, 105, 121

batch melting, 150, 163, 166, 167, 169
batch crystallisation, 55, 69, 71
Bruderheim meteorite, 8

chemical analysis, methods, 10
closed system model, 215
coefficient of variation, 3
compatible elements, 58
complex formation, 33
constant melt proportion, 68
constant mass increments, 7
continuous melting, 212, 213, 218
crystal nucleation, 51

degassing, 87, 91
diabase, W-1, 3
differentiation index, 125
disequilibrium melting, 179
distribution coefficient, 18
Doerner–Hoskins process, 61, 62, 72
dynamic melting, 211, 213, 239

early partial melting (EPM), 14
element ratio plots, 133
equilibrium
    constant (K), 21
    crystallisation, 51, 53, 129, 130
    melting, 155, 158, 159, 168
    partition, 25
Eu anomaly, 24, 30
exsolution, 87

fluid
phases, 86

release, 89
fractional
    crystallisation, 51, 55, 65, 91, 130, 131
    fusion, 146
    melting, 152, 155, 157, 159, 168
fractionation efficiency, 63

garnet
    lherzolite, 192
    peridotite, 193
Geostandards Newsletter, 5
Gibbs free energy, 20
granite G-1, 3
granitic rocks, 2
granodiorite, 7
growth rate, 51

Harker diagram, 124
Henry's Law, 43, 44, 46
high field strength elements (HFS), 58
hygromagmatophile elements (HM), 59

incompatible elements, 58
incongruent melting, 94, 187, 188
incremental
    batch melting, 160
    crystallisation, 66, 67
intercumulus trapped liquid, 63
interfacial energy minimalisation, 182
inversion modelling, 135

kinetic disequilibrium, 74

large ion lithophile elements (LIL), 58
lattice deformation model, 38
    law of mass action, 17
leucite–anorthite–silica system, 76
liquid line of descent, 124
liquidus, 53
lunar soil 12070, 5

magma
    recharge, 105

system, 52
major elements, 1, 83
  effects, 27
mass
  balance, 95
  porosity, 239
melt
  dynamics, 232
  percolation, 237
  structure effects, 34
melting, 142ff.
  with volatiles, 170
mesostasis, 63
mineral
  chemistry, 39
  pairs, 66
  zonation, 61
minor elements, 1
molar partition (Nernst) coefficient, 16

Nernst coefficient, 16
network
  formers, 34
  modifiers, 34
non-steady state, 120

olivine–melt partition, 27, 39
Onuma diagrams, 35, 39
open system
  crystallisation, 120
  melting, 220, 228, 231, 239

partition coefficients, ix, 14, 32, 36, 37, 42, 54, 201
  and Ca content, 40
  and ionic radius, 37, 38, 41
  apparent, 62, 66
  bulk, 150, 202
  measurement, 23
  tungsten, 31
  use, 44, 45

values, 151, 219
  variation in, 78, 202, 223
Pearce element ratio (PER), 126, 127
percolation of melt, 237
peridotite melting, 156
peritectic reaction, 54
permeability threshold, 212, 239
phenocrysts, 6
plagioclase binary system, 54
precision, 3
premelting, 148
pressure change effects, 22

Raleigh fractionation, 75, 79, 91, 92, 131
rate process model, 116
reaction melting, 187
reactive melting, 194
redox effects, 29
resorption, 94
residual magma, conservation, 112
retrograde boiling, 87, 91

sample heterogeneoity, 5
sampling, 9
Sedona conference, 25, 45
standard reference materials, 4
substitution site deformation, 34

trace element path (TEP), 136, 138, 139
trace elements, 1, 78, 152
trapped liquid, 64

variation diagrams, 124
vesiculation, 87
volatile
  components, 87, 171, 173
  fluids, 31

water concentration, 34

Zone refining, 205, 206